A PRIMER ON ELECTRONIC SECURITY FOR SCHOOLS, UNIVERSITIES, & INSTITUTIONS

SECOND EDITION

By:

Henry L. Homrighaus Jr., PSNA, CHS-V

Frank J. Davies, CVI, CIPS, CHS-III

Gregory Bernardo, CVI, CSI, CHS-I

Printing History
Second edition published March 2012
First edition published December 2005

 For information address: Aella Consulting Group, Inc., P.O. Box 950, Argyle, TX 76226

ISBN: 978-0-9853731-2-2

WHAT IS NEW IN THE SECOND EDITION?

The success of the first edition primer by Henry Homrighaus, Jr. and continued interest has validated the need for an updated Second edition. Frank J. Davies and Gregory Bernardo as recognized Subject Matter Experts in physical security have joined Henry to write this completely revised Second Edition. They bring a wealth of diverse industry experience and were invited to add to and round out the technical flavor of this edition. Since the publication of the first edition, the security industry has witnessed many substantial changes, including significant maturity of digital video products, and fast progression of IT and IP based hardware.

Depth and detail is enhanced in all sections of this edition, and new subject matter has been added in the following areas:

- Perimeter Protection and CPTED - Crime Prevention Through Environmental Design
- Physical Security Information Management (PSIM) Systems
- Mass Notification
- Video Surveillance including IP Video and Digital Video Management System
- Video Analytics
- Visitor Management
- School Bus Safety
- IP Based Access Control
- IT infrastructure
- Wireless Transmission Methods and Devices
- Cloud Computing and Cloud Storage
- Budgeting
- Maintenance and Inspections
- Standards and Certifications
- Where to Get Help
- Access to Online Project Checklists & Resources

FOREWORD

There are numerous factors driving the need for meaningfully increases in securing our children's' schools, not the least of which are the recent eruptions of violent events that have caught schools or officials off guard or unaware of the potential threat.

The educational process affects all of us in so many ways. Learning is the one thing we do every day of our lives. Teachers and educators formalize this process and are responsible for motivating many of us in our formative years.

Clients usually contact us to provide subject matter expertise on the selection and use of security technologies. In most cases, our first task is to educate the client that the technology does not work by itself.

This book offers an overview of security projects and practical guidance to make the reader better informed about decisions on security technology. This book does not replace the use of appropriate expert advice, but it does introduce many of the key steps and technologies we consider in developing physical security plans for our clients. Our primer also acquaints the reader with considerations involved in planning, selection, budgeting, funding, bidding, installing, training/operating, and managing the security process in their school, university, or institutions.

WHAT WE COVER INSIDE:

- The need to mitigate the impact of school tragedies and deter new tragedies in school districts, universities, and educational campus environments.
- The need to build a team (not a committee) and *who* should be on that team.
- Considerations when building your security plans and budget.
- The application of available security technology.
- An understanding of how to anticipate and minimize recurring security maintenance costs.
- Understanding and evaluating the rapid advances in physical security technology.
- Purchasing methods and practices are discussed to examine and ensure that the best systems and equipment are purchased for the each application.
- Contract vehicles and ways they are improved to ensure quality bid responses.

"Regret for the things we did can be tempered by time; It is regret for the things that we did not do that is inconsolable."
Quote from syndicated
early 1900's journalist Sidney J Harris,
Chicago Daily News & Chicago Sun Times

TABLE OF CONTENTS

CHAPTER 1 - THE GROWING SECURITY RESPONSIBILITY FOR EDUCATORS

Today's security concerns are so many and so varied that institutions have difficulty coming to grips with all of them. This fact is not surprising; given their reason for existence is education and not security.

With parents leaving their children in the educator's care, concern for student safety has always been an assumed responsibility. However, what used to be a simpler responsibility of making sure children were transported and educated has evolved to include a full time doorstep-to-doorstep "security".

The expectations for security and safety in our educational facilities have grown considerably in recent years due to significant societal changes:

- High percentage of families with both parents working fulltime.
- High divorce rates where equal parental custody can no longer be assumed.
- Increasingly litigious actions by parents and others regarding school related incidents.
- Educational Employee workplace safety.
- Validation and verification of visitors and the "real risks".
- Local, Regional and Global threats of terrorism as a very real concern.

Vandalism, graffiti, burglary, arson, contraband, student violence, assault, sexual assault, rape, extortion, murder, drugs, bullying, physical threats and violence against staff and students are security issues that administrators are forced to deal with all too often.

Many schools are self-insured to a great degree, and the loss of a school facility could result in the inability to handle students, the possible loss of funding, and the requirement for more bonds and taxes to replace the facility. It is a false sense of economy not to adequately protect our schools from the devastating effects of fire, safety, or other liabilities and security risks.

To function in this environment and enjoy as much freedom as possible, parents now expect a much higher level of overall safety & security. Parents want and need to feel confident that the educators have everything under control (with a broad definition of everything).

That is a tall order for school administrators given the increasing demands on the educational institutions and their resources. Today costs continue to rise, continued growth and consolidation is now the norm; and state and federal government regulations seem to change almost daily. Funding through bond elections, school board changes and actions, and technology changes all complicate a key element of a sound security plan. Lastly, after a security plan is implemented and technology is installed, there are maintenance costs and responsibilities often forgotten or overlooked in the process.

There are solutions to most problems if school administrators are aware of them and knowledgeable about the resources at their disposal.

CHAPTER 2 – BUILD THE TEAM

Managing all of the aspects of the security process can be a daunting challenge at best. There are so many variables to understand and deal with in giving some definition, operational guidelines, and compliance to your security program. This is amplified by all of the different circumstances found in schools, universities, and institutions. Due to size, budget, and location and you often have a frustratingly complex issue to control.

Larger school districts and university systems have the luxury of dedicated professionals at the helm of their various departments. These are typically professionals in procurement, law enforcement, information technologies, and engineering with a budget and staff in place to handle these processes. Having said that, one of the truisms of life is that there is never enough budget or staff to handle all the situations or problems all the time.

In smaller facilities, educators, principals and superintendents often find themselves doing double duty with little guidance or direction Educators today have so many guidelines, requirements, and mandatory procedures thrust upon them by the system that it is difficult to handle addition assignments no matter how necessary or well intentioned. It is our intention and expectation that this book will be of assistance to most of you.

As an initial step, a security team should be created for the institution (if not already existing) with each member assigned specific responsibilities and authority on the team.

We suggest a team with at least one representative from:

- the district and/or school administration,
- local police authority,
- internal security,
- the Information Technology (IT) department.

Notice we suggest a "Team", not a committee. A committee designation is typically associated with a loosely coupled grouping of individual's assigned tasks with no significant responsibility or authority. However, a Team conveys common, combined goals and strategies to effectively and decisively deal with the focused issue.

It is essential that the team be selected with sufficient knowledge of the institution and the issues, so they can aggressively pursue sources of information. This team must have the authority to request, search for and collect all information required. This empowered team will be the oversight through the entire process and provide critical and key decisions that affect the short and long-term success of the plan for their school, university, district, or institution. Many of these team members should be the keepers or custodians of the finished product; to insure they have a vested interest in the outcome.

The natural divisions in most educational institutions allow for the assignment of many segments of the security management process to the departments best equipped to deal with them.

- Administration is the hub for all of these activities directing traffic so that all of the necessary functions are assigned and handled to insure completion and compliance. This is the natural scheme of things and works well in most cases.
- Procurement or purchasing normally deals with contracts for service, monitoring, maintenance, and inspection. They arrange to pay for the services as rendered and give advice to administration when it is necessary to renew or acquire new quotes for services as contracts expire. They will help the project to keep an eye on finances, but they generally do not have the experience to manage a security project.
- Information technology is a resource to manage we rely on to monitor or transmit information from our various security systems. In almost every new security system, the facility's network is the backbone. IT should always be involved in planning any network applications concerning physical security systems as they are often left out. Most state of the art security systems can be managed over your LAN/WAN. Many IT departments are responsible for data and cabling installation, upkeep, and as well as maintenance.
- Engineering or building maintenance in your institution is a very critical part of the overall management plan. This group typically handles all of the electrical, locks, and building maintenance in a school or district. Engineering is can also be charged with the ongoing maintenance and inspections required by law in many jurisdictions where your fire alarm systems are involved. It is essential to involve them from the beginning of the design process, as they will inherit the support for much of the physical hardware installation of your system.
- If your school or university is large enough to require its own law enforcement and patrol officers, you are in a unique position to control the security management process. In many cases, local law enforcement and security personnel also serve as the security system monitoring personnel. If you do not have your own onsite law enforcement, the responsibility will often fall to the onsite principals, assistant principals or other staff members. In either case, you will want to have the input of a local

authority because they live with the onsite conditions and culture every day.

One or more members from of each of these departments should commit to meet at least once a month (more often if just starting your project) at an established time dedicated only to the security management process.

Your team will need to formulate a plan with all the individual components required to implement an effective security solution. Whereas your team may include some truly motivated individuals, team meetings should be the primary forum for far reaching decisions regarding assessments, planning, equipment evaluation, preparation of offerings, and any other items dealing with the overall security of the institution.

It is also a good idea to hold off on vendor demonstrations until after your team has identified your institution's primary needs and evaluate each vendor against those needs. It is very easy for a decision maker to become enamored with a product that does not ultimately fit the institution's overall needs.

At the initial meeting, the team must:

- **Select a Team Leader**: The team leader is the individual who can commit the time, have access to resources, and be able to assist the team in achieving its ultimate goals. The team leader should also be a champion to the project, and someone who can exert some degree of influence should the need arise.
- **Define the Overall Objectives**: The key in defining your objectives is not to focus on costs at first, but to focus on the ultimate requirements, goals, and solutions. Everyone should have input, but the motivation is safety and security for the entire institution, not the creation of who will control what.
- **Establish a Schedule**: The Team, to be effective, must work with a sense of urgency. Establish key milestones for each step in the process and a timeline for which these are made.
- **Assign Responsibilities**: One individual cannot perform or be responsible for all tasks. Assign key tasks to team members. The team Leader should then focus on the actions required to achieve the tasks.
- **Follow-up**: Make sure there is consistent follow-up on each task. Do not wait until the due date to find nothing has been done.

- **Document:** Establish a documentation process to ensure that every aspect of the planning process is documented to support the ultimate recommendations. The results of the process will be subject to administrative and more than likely public scrutiny. Therefore, each step the team makes must be supported, substantiated, and documented in every way possible.
- **Revisit the Objectives**: Throughout the course of plan development and implementation, new ideas or items that are overlooked initially need to be identified. Be flexible and adjust the plan to achieve the maximum results.

The realization that actions are required by an established date is the best motivator to achieve the objectives. Identify a target date for every task and effort. Team members should know that they would hold up the project or other members of the team if they were late to deliver.

CHAPTER 3 – ASSESS NEEDS

The team must formulate a plan and policy for gathering, evaluating, and using information to produce an assessment of your own unique security issues.

The success of the planning process is directly dependent upon the team's ability to identify and procure/secure data at this point. They need to know just what is happening in the community, what types of crimes are being experienced and with what frequency. This information will create an understanding and appreciation for the magnitude of the existing problems or the potential for problems.

Once the issues are identified, the team can prepare a plan that will begin to direct existing resources, identify new current needs, and plan for future requirements. This may require the reassessment of needs for additional funding.

At this point, we still have not mentioned security technology solutions and your team should be cautioned not to jump into technology evaluations until they have completed the assessment process.

Risk & Needs Assessments

A prison is understandably a singular purpose built facility and is designed by architects and engineers who specialize in prisons. Unfortunately, this is not the case for schools as they serve a multitude of services to the community, not just as a school. For this reason, schools have a wide range of hurdles to overcome when considering their safety and security plans.

Risk assessment is the first step in the risk management procedure. Risk assessment is the task of determining the quantity and size value of risks related to each situation and the recognized threats or hazards your institution may be faced with.

Risk assessment is an objective evaluation of risk by a skilled individual or individuals. The performance of the assessment should account for all likelihoods and uncertainties and present the conclusions in a clear and well-documented form. Risk management and its initial assessment are difficult and ongoing tasks over the life of an institution. Placing the correct value and potential for occurrence is the key to assigning a protective measure and cost to your management of risks.

A needs assessment is an important part of the planning process. Needs assessment is a logical step-by-step process for determining and addressing your institution's needs, or gaps between conditions that you identified in your risk assessment, and the identified shortcomings of your current conditions (outside of the risk assessment, these are your needs or wants). The list of needs you identify can be a combination of identified deficiencies and the desire to improve current performance.

With the risks and needs identified, your finite resources can be directed towards developing and implementing applicable solutions. This data gathering and analysis needs to be appropriate and sufficient for your facility's environment in order to effectively address your institution.

Older Schools Were Not Built To Address Today's Security Concerns

Recognize that today's security issues are significantly different from when facilities were initially built. Many of our country's existing schools were designed as common commercial building and not purpose built with security in mind. Older schools are often "grandfathered" into older building codes and may not have been required to install the systems and infrastructure of a newer building right next-door. It is important to look for and minimize these "dangling" risks and exposures as part of your overall plan.

Permanent "Temporary" Prefabricated Buildings (Prefabs)

In recent years, schools have struggled to meet the demands of a growing student population under tight budget constraints. An unfortunate trend is the use of generic temporary prefabricated buildings, which ultimately become permanent. These structures are typically not designed with a consideration for school security and are often provided by the lowest bidder. Risks with these types of structures are higher on a per square foot basis than standard structures.

Newer School Designs Still Lacking

Many of today's newer school designs are unquestionably "purpose built" and give our children every chance for a quality education. However, even in these newer facilities security is often one of the last considerations (and one of the first to be cut if the budget is tight). If the design requires only the minimum legal system requirements, then that is exactly what the lowest bidder will give you.

It is essential that all aspects of your security program be examined, any weaknesses identified, and any existing security measures noted. Security assessments should be performed on a site-by-site basis, ***without assumptions***, and then summarized institution-wide to identify trends and common issues.

Assumptions (especially unspoken ones) are one of the most common causes for change orders and project overruns!

Due to the enormity of the task, educational institutions and businesses will often look to outside consultants for Site Vulnerability Assessments or Site Security Assessments. However, an appropriately staffed and motivated team can do some of it themselves. It is always wise to have a professional validate the team's results.

Other resources to help with these sites assessments can be found in building code departments in most cities and states that can be used to develop a survey or report out of compliance issues. Most city, town, or institutional employees have a stake in protecting schools and insuring the education of their citizens. The local fire department can survey our existing fire protection and make suggestions. The crime prevention specialist from the sheriff or police departments can survey your schools and provide suggestions.

Insurance carriers will occasionally make their services available to clients (or prospective clients) by doing surveys for risk analysis. However, care should be taken using their analysis because their objectives are often different from yours.

Assess Your Existing Security

During the initial needs assessment period, the Team should identify existing security measures and ascertain who is in charge of what. In many cases, this will be prove to be enlightening and will provide the Team with an opportunity to assess the status and validity of the measures/systems currently being utilized. This will also help them define the scope of responsibility for particular segments of security, and formulate recommendations for those functions in a final analysis document.

Ultimately, the institutional security policy should be defined and written in a security manual that is made available to all parties on an as needed basis. If the institution does not have a security manual, then this should be high on your list of priorities. This manual should be the framework of how, who, and in what manner every level of security and safety is handled in each institution, and they need to be site specific.

Determining who is best equipped to manage individual security components should be part of the team's assessment. One of the best approaches is one in which a single individual is in charge with a "staff" (full or part time) of responsible personnel for each aspect of the security policy. In larger institutions, Memorandums of Understanding (MOU) may be in place between departments identifying shared security components, reporting, or systems that are installed by one department and maintained by another.

The final part of the assessment of existing security measures is to "inventory" all existing security tools, systems, staffing, policies, and procedures. This is an important step in understanding the status of the institution. The team should start by acquiring up to date floor plans of each building and inserting symbols to indicate the location, type, and coverage of all installed security devices.

A facility-by-facility checklist is needed to determine exactly what is being done or utilized in each facility. Some of the components of the checklist are listed below and available online as described in Appendix A.

Inventory of Existing Electronic Security Components

The team should start with the basics (Some of these seem to be common sense items, but you would be surprised.):

- Are all locks working correctly?
- What type of Security systems currently exist?
- Are they owned or leased?
- Are they in operation?
- Are they monitored, if so by whom?
- Have they proven effective and what are the operating costs?
- Is there a test and maintenance plan and how does it handle replacing obsolete equipment
- What is the LAN/WAN Network capacity in this location?
- What is the capacity and type of local support at this site?
- How critical is coverage in the identified locations (simple scale)
- Is the equipment tied to a response plan, policy or procedure (identify)
- What is the exact coverage of a camera or motion detector?
- Are there dead spots in coverage?

The above list addresses basic security system components (locks, etc.). A similar checklist should be developed and implemented for fire alarm systems, CCTV (Closed circuit television), and any other security systems in place. All components of any existing security systems should have been documented at installation (referred to as "as-builts").

With the fire alarm systems, these should be documented on the original drawings per the original installation specification or code requirements. Most states require that fire alarm plans be submitted to the local authority, i.e., Fire Marshall, Fire Chief, City Planning departments for approval and certification with a record copy on file.

Bring the Big Picture into Focus

Given all of the information you have now collected, the Team should be able to see how well (or not so well) facilities are protected. It is now time for the team to begin the move forward in development of their security plan.

The initial results of an institution-wide assessment can be a rude awakening to many school administrations and can bring a sense of security being an overwhelming and insurmountable task. However, the team can temper this with assurances that the final analysis is yet to come. This analysis will result in a more detailed security plan that will address systems, financial, and implementation requirements to insure the safety and well-being of the faculty, staff and students in the facilities.

CHAPTER 4 - DEVELOP THE PLAN

Security systems do an excellent job of protecting buildings and contents when properly designed and installed correctly according to the manufacturers guidelines and instructions. The best designs employ a multilayer approach in which there are ties between primary systems and redundancy or overlap is created within the design approach. We recommend a hierarchical project design, which utilizes subsystems that can operate stand-alone in the event of failure of another system.

For example, many access control systems offer the inclusion of video. However, if the access control system fails and it is the control for video, then the video also fails. A better design would be an integrated access control system with a separate video system head-end in which they are tied together via a network. When the two systems are tied together via an exchange of data, they can be operated together or mutually exclusive in the event of failure.

Chapter 4 offers a brief overview of topics, methodologies, and commercially available technologies valid for physical security planning in most institutions. In Chapter 5, we will give you a more detailed look at security technologies we usually consider in developing physical security plans for our clients.

Time Considerations

It is important that you allow sufficient time to fully develop and initiate your plan. Some of the more time consuming efforts will include:

- Fully investigating current systems
- Determining the needs of your institution
- Assessing available technologies
- Developing a specification to meet your needs.
- Developing a full and comprehensive plan on how the technology will be used in your overall approach.

Considering these tasks, the amount of time required is solely dependent on the size, scope, and number of locations that you will be securing under this particular endeavor.

Tiered Physical Security Systems Approach

In the following chapters, we refer to a "Tiered" method of constructing your physical security system. This method of creating multiple layers in a hierarchal group of systems, and how they should react to provide levels of autonomy, redundancy, and backup is a method proven repeatedly to work.

The tiered hierarchy approach of physical security places the highest (most overarching product) as Tier 1 and subsequent lower level systems further down (Tier 2, 3, and so on).

Some systems such as Fire and Intrusion can be Tier 2 or Tier 3 depending on how you apply them to your project design.

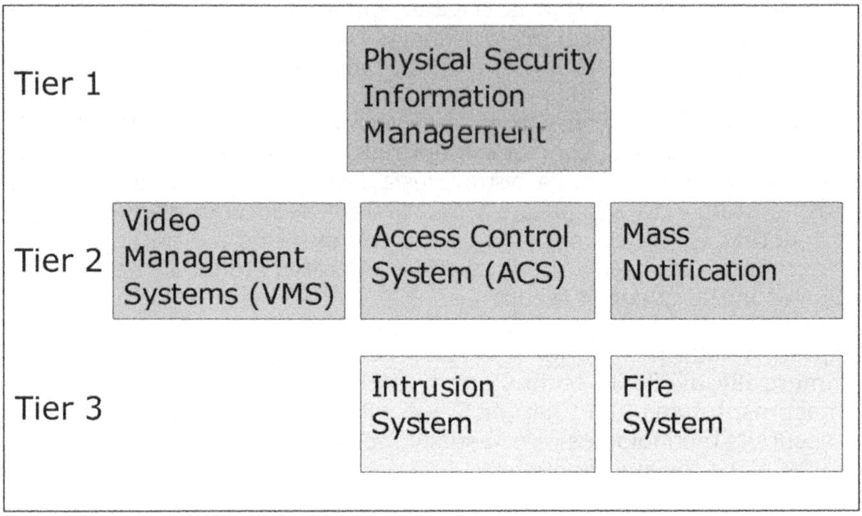

The above example shows a Tier 1 PSIM system (discussed later in this chapter), which can receive information from Tier 2 systems such as access control, video management, and mass notification. This example also shows Tier 3 systems, Intrusion and Fire, connected to Tier 2 systems. All five of the tiered systems work autonomously and each system can be addressed directly or through the next higher tier.

A good example of autonomous tiers can be a fire system and a PSIM. Security guards using the PSIM can be informed about and acknowledge active fire alarms. However, in most cases someone has to go to an alarm panel to disarm an actual alarm.

The Basics of a Physical Security Plan

Physical security plans, when properly planned and implemented, can serve a host of functions very efficiently. Only careful planning and realistic expectations with a thorough understanding of what is possible will insure the proper deployment of security systems.

The plan your team develops to mitigate the issues found in the previous assessment stage should include the following elements - deterrence, detection, delay, response, recovery, re-evaluate.

- **Deterrence** includes policies, procedures, and security devices/systems to defend against attacks on people and property.
- **Detection** - includes monitoring for unwanted acts by personnel and/or technology.
- **Delay –** includes measures to delays intruders long enough to allow a security team to apprehend them before they achieve their objective or escape.
- **Response** –includes procedures and actions for responding to a security incident in progress.
- **Recovery** - includes procedures and actions to occur following an incident. This could be as straightforward as filing a police report/insurance claim, or it may involve contacting parents and legal counsel.
- **Re-evaluate** - Like all plans, physical security plans are no different in needing continuing periodic updates. Plan on regular updates to account for new knowledge base information, changes to the school, changes in the environment surrounding the school, and most importantly changes within the populous attending and managing the school.

Additional information regarding these concepts can be found in numerous books and training programs (see Appendix A). If you do not have the time to dedicate to become an expert in physical security mitigation (or you have a job description with other responsibilities), you can rely on security consultants like the ones found on http://www.ConsultantRegistry.org.

Effective Policies and Prudent Use of Technology

The initial priority of many institutions is to attempt to keep problems from reaching their facilities using an "outside-in" approach. Under this approach, the first areas to address are external to the buildings such as parking lots, playgrounds, approach roads, fields, unattached structures, etc.

Consider all your options when mitigating an issue identified in the assessment stage. Risk mitigation can include various combinations of added manpower, security, or law enforcement presence, physical changes to the area, and adding security technology.

Sometimes just the presence of some security products (like cameras) can prevent incidents from occurring. In some cases, changing a policy about which doors are open during school hours can be more successful than adding a security camera.

Recognize that most security technologies have some form of associated maintenance costs that should be considered in addition to the initial purchase. We discuss the concept of total cost of ownership in a later chapter.

Emergency Response Planning (ERP)

Regulatory requirements and guidelines for Emergency Response Planning (ERP) are becoming more prevalent, but are far from universally applied. Where they exist, these guidelines differ district to district. Some states have enacted regulation; others have established guidelines, and in some offer no guidance.

Federal Agencies currently provide some guidance and template assistance in the items that they feel are important for from a federal level (See appendix A).

Federal funding sources for emergency planning have been discussed, promised, and then canceled numerous times in the past decade. Where federal funds do exist, we find them to be grossly inadequate and not widely available. In order to obtain federal funds, we recommend you contact a grant writer. We offer links regarding funding sources through our information portal described in Appendix A.

Though not directly applicable to ERP or physical security, many schools have substantially improved their network infrastructure using E-rate program grants funded by the Universal Service Fund (look for the fee on your phone bill). While these grants cannot directly be used for security upgrades, network based security systems can benefit from the resultant improved network upgrades and internet connectivity. School officials interested in E-rate grants should act quickly due to the expected discontinuation of the Universal Service Fund in 2017.

While this book primarily focuses on security technology, the authors wish to stress that a specific and dedicated Emergency Response Plan for your school is one of the most important items that school administrators can do to provide for the safety of the entire school population. Emergency Response Planning is a specialized effort, and skilled profession help is highly recommended in this effort to limit the liability of the school administration, but more importantly to provide all of the varying scenarios that should be planned. Please see some of the additional resources in the appendix area of this book.

Crime Prevention Through Environmental Design (CPTED)

CPTED is way of mitigating security risks identified in during your assessment. CPTED strategies emphasize enhancing the perceived risk of detection and apprehension. The three most common environment strategies are natural surveillance, natural access control, and natural territorial reinforcement.

Natural Surveillance increases the threat of apprehension by taking steps to increase the perception that people can be seen. For example, when creating lighting design, avoid poorly placed lights that create blind spots for potential observers and miss critical areas. Ensure potential problem areas are well lit: pathways, stairs, entrances/exits, parking areas, etc.

Natural Access Control limits the opportunity for crime by taking steps to differentiate between public space and private space. For example, use a single, clearly identifiable, point of entry, use structures to divert persons to reception areas, and use low, thorny bushes beneath ground level windows.

Natural Territorial Reinforcement promotes social control through increased definition of space and improved proprietary concern. Within an institutional setting, placing amenities such as seating or refreshments in common areas helps to attract larger numbers of desired users. Scheduling activities in common areas increases proper use, attracts more people and increases the perception that these areas are controlled.

In terms of effectiveness, a more accurate title for the strategy would be crime deterrence through environmental design. Research demonstrates that CPTED cannot literally prevent offenders from committing crimes. Retrofitting an existing environment to meet CPTED can sometimes be costly, but when incorporated in the original design phase of facility planning, cost of designing to CPTED principles are often lower than with traditional approaches.

Planning Guidelines

There are several rules of thumb in designing and planning for the use of these systems: Visualization is extremely important in the planning. Develop your layout and design using scaled and accurate floor plans.

Use standard and consistent symbols to represent each device in a security system design. (There are several symbols libraries available - See Appendix A)

Insure that the legend explaining the device symbols is included in each drawing set. Do not assume that anyone will automatically understand your intent.

Be detailed and precise, these planning documents will be relied on to create your bid documents and build your system.

The use of Building Information Modeling (BIM) has made steady and significant growth in the construction market. BIM is still an emerging standard in the security industry. Find out if your facilities / engineering department requires BIM before you start your project.

Which System is in Control and Who will Monitor all the Systems

For the remainder of chapter 4, we will introduce you to the technology subsystems that we rely on for physical security. Historically each of these subsystems has been developed as an independent system by unrelated manufacturers. For this reason, most subsystems are designed to run independently and interface with other lower tier sub systems. For this reason, security technology can be implemented in phases, adding one or more complete subsystems at a time.

As you evaluate subsystems, be aware that most manufacturers will encourage you to use their system as the head end. For example, many access control system manufacturers will tell you that they can also be your interface for controlling a video system (and some of them can). However, each manufacturer can only afford to interface with a handful of other products.

Relying on one subsystem to control another subsystem may be perfectly adequate for small to medium size facilities. However, doing this on larger facilities or campus environments can paint you into a very expensive corner. Beyond the cost to change out or reconfigure equipment, operators who have been using the original configuration will be interrupted and forced to learn to operate the new system(s).

In many medium and large applications, we recommend the use of a PSIM (Physical Security Information Management) or command and control system, which all the subsystems will connect.

Physical Security Information Management (PSIM) / Command and Control

In mature installations, all of the individual technologies eventually consolidate into a common operating picture. Getting to this stage is usually the target; however most institutions usually get there piece by piece.

We start with "Monitoring all the systems" so you can avoid frustration and wasted money because parts you chose early on don't work as your physical security installation "grows up".

Before looking at Tier 2 systems like video control systems and access control system components, determine how your system(s) will be used and who will use it. Here are some considerations:

- Will the onsite principal be the primary responder? Do you need a secondary responder?
- Will you (now or in the future) have dedicated security personnel or law enforcement presence?
- Do you need to view site conditions offsite (alarms or video)?
- Consider outside influences like weather conditions prohibiting onsite personnel.
- Allow remote police logon during events.
- Anticipate visual confirmation of alarms to reduce false alarms or fees.
- Do you have public gathering areas for sports events or other after hour events?
- Do you have problem traffic areas (people or vehicular)?

A well thought plan provides a true operational picture of all systems that impact security and life safety. By using a PSIM software platform, your disparate systems and those that have been brought in under your planned expansion can provide you with usable information to aid in the decision-making and responses to all types of events.

Security Personnel

Depending on the size of the institution, long term security plans usually include a combination of dedicated institutional security staff with appropriate authority, guard service and/or law enforcement. Security personnel are an absolute necessity for the response portions of your security plan where the ability to make intelligent situation based decisions is necessary.

Often we see facilities start by positioning existing educational staff at critical locations to monitor activity and keep turmoil to a dull roar. These staff members were not originally hired to do security but become part of a short-term security plan because it is something the institution can do immediately.

This practice can only be considered a short-term solution because it creates a strain on educational staff. If the institution is not large enough or not ready to deploy dedicated security staff, then the next likely action will be to shift this workload to third party security guards.

Logically, in both these cases, your security people cannot be everywhere all the time. This is where security technology gets applied to work hand-in-hand with your security personnel. The application of technology in this case is called a "force multiplier" because the technology enables you to do more with less people.

Consider that your long-term physical security plans ***and your budget*** will dictate whether your security personnel provide a reactionary role (that handles incidents after they occur), or a significant proactive role (which will reduce incidents thereby improving overall security and safety).

The decision to become proactive means a larger investment in time and money. Ultimately, your security plan will need to include the right mix for your own unique environment.

Centralized Dispatch / Central Station

Fire systems & intrusion detection systems are purely reactionary in nature. You will not hear from the systems unless something is wrong. For example, smoke is detected, or a window is opened after hours.

The reason for using technology in your security plan is so you do not have the ongoing cost of paying a large army to protect your facility. *Depending on the size of your facility, you may need a small army*. Under the same logic, you would not want to pay someone to sit next to every alarm panel and then call the police or fire department.

One consideration is utilizing a centralized dispatch or "central station", which can monitor your security technology systems. Often, monitoring will be performed offsite by a central station provider that is certified to monitor fire and life safety systems.

Offsite (third party) central stations are limited to what systems they can interface with. The limit of what services they can offer is primarily due to technical challenges resulting from a lack of technical standards in the security industry. Recently, central stations have started offering to remotely connect to your video system to confirm an intrusion and reduce false alarms.

If your school or university is large enough to justify a central station or monitoring center, then you can significantly leverage security technology to improve security and safety especially after hours.

For example, onsite operators can:

- Monitor areas by video for problems where people congregate (other potential benefits of intelligent video also known as Video Analytics will be described in Chapter 5).
- Remotely open doors for staff that have misplaced their access cards or keys.
- Audit security records to determine who accessed a secure area following an incident or attempted to access a secure area they were not authorized.

How your systems are monitored and insuring proper authorities are dispatched according to the nature an incident is a critical part of your security plan.

Mass Notification / Sounding the Alarm

Mass Notification has evolved to become a comprehensive solution that often includes software and cutting-edge communications technology to warn people of danger and keep them informed. Administrators, Security and Facility managers who have not started developing mass notification plans should be aware that they will soon be compelled to implement a system. Mass Notification is being addressed by a greater number of local codes and standards and recently became mandated on the federal level under the Clery Act.

Another significant development on the code front is the National Fire Protection Association's new fire alarm and signaling code, which includes expanded information on mass notification systems. In its 2010 edition (NFPA 72-2010), the 2007 "Annex E" guidelines on mass notification were replaced by a more comprehensive standard, and the new chapter on Emergency Communication Systems has a broader scope.

The new code includes requirements for three different types of mass notification systems and also contains a directive for two-way radio enhancement systems, which will help fire service personnel coming into a building. Whereas visual and audio-based systems are straightforward to implement, two-way radio enhanced systems require additional skills because steel and other building materials attenuate communications.

The new code tends to be quite detailed. Extensive requirements for risk assessment have also been added. As mentioned above, the NFPA requirements are becoming a standard. These standards have already been adopted by federal directive in military and defense buildings.

Your physical security plan needs to address how you alert your security staff, faculty, students, parents, and the public.

Physical Security Technology Systems / Sub-Systems

Up to this point, we have discussed several long-term considerations that you should build into your overall security plan. Now we will discuss the individual subsystems, which are considered the backbone of every physical security technology plan. As we mentioned above, remember that each of these systems can often be installed independent of other system. This practice needs to be balanced with your long-term needs to avoid painting yourself into a corner with a particular technology or brand.

Video Surveillance / CCTV / Digital Video / Video Management Systems / Cameras / Digital Video Analysis

Video Surveillance, or the use of video monitoring within educational institutions, has become standard practice to provide safety, liability protection, and forensic evaluation for incidents or events that may occur. As recently as ten years ago, many schools wrestled with decisions about its use and justification. The benefits of surveillance cameras have proven to outweigh the associated cost. Today most parents are comfortable with cameras, and many times expect to see them in their children's schools.

Video surveillance as a technology continues to mature and is one of the fastest changing components of the security industry today. For that reason, you may hear references to CCTV, Digital Video, Video Surveillance, Security Cameras, Video Management Systems, and the Digital Video Analytics. However, these terms are not necessarily interchangeable nor do they refer to the same system type.

Video surveillance systems that we see installed in educational facilities are comprised of a Video Management System Platform that interacts and controls video surveillance components (such as cameras) and other video related technology such as Video Analytics. This chapter will outline many of these products and describe how they interact. Available and recommended technologies are described in more detail in Chapter 5.

Video Management System (VMS) Platform

A Video Management System (VMS) can be considered either a comprehensive software core which interacts and administers all other video surveillance components, or it is a key ingredient in a suite of security/life safety components.

Regardless of this distinction, most veterans of the field would agree that not all VMS are created equal.

Most VMS will at least:

- Provide a user interface
- Administer users
- Manage and facilitate the transport/recording/playback of video
- Offer some transport and presentation of alarms
- Perform some level of activity logging

Manufacturers in the video surveillance industry routinely advertise their products with the moniker "Enterprise" though it has little or no correlation to this predominately IT term. We cannot point you to a reliable scale of quality for VMS, because there is no recognized certifying body for this industry.

The best way to communicate the benefits of one VMS over another is to give you an idea of the features we look for when designing a system for educational facilities. In addition to a study of the manufacturer and a product's long-term viability, we look for scalability and comprehensive feature sets, which a client can grow into or expand upon. Each VMS system contains a varying number of the following features typically applicable to educational facilities.

Some of the Scalable functionality we look for includes the ability to:

- Manage a large number of video devices
- Interface with Open Source cameras and video encoders
- Interface with other security industry sub-systems (access control, intrusion, fire, etc.)
- Interface with Physical Security Information Management (PSIM)
- Manage a large number of disparate users
- Utilize Network Attached Storage or Storage Area Networks

Mature features we look for include the ability to support:

- Diverse user groups with flexible permissions
- Business rules applicable to devices and 3rd party interfaces
- Comprehensive audit trail (both at the user level and administrator level)
- Built-in video motion detection and video analytics (described below)
- System health monitoring

Other video surveillance components we alluded to above are described on the next pages.

Cameras & Camera Placement

There are two types of security cameras; fixed position, which you use to watch one area, and motorized Pan-Tilt-Zoom (PTZ) cameras, which can investigate a much wider area. An operator can control a PTZ camera, or it can move automatically in response to an alarm such as a door being forced open (if you have the intrusion system interfaced to your video system).

Just like consumer digital cameras and today's smart phones, security cameras come with different fitments. Surveillance cameras are available in many resolutions, with different lenses, and with various low light level characteristics. Recognize that higher resolution brings additional technical considerations that we will talk about later in Chapter 5. Security cameras are also available with special features (including cameras that can see at night) and styles to fit most applications and aesthetic concerns.

Most institutions strongly favor the use of cameras inside dome housings (dome cameras) for aesthetics and the fact that domes make it difficult to tell exactly where the camera is pointing thereby deterring a potential incident.

Camera Placement

Your video recording system can only record on a camera's primary view. Consider how your eye works when you are driving and a bug smashes on your windshield. When you focus on the smashed bug, everything further away is blurry. Security cameras are similar in this regard and to be effective, cameras need to be placed, aimed, and focused with specific intent.

Be wary of claims that pan/tilt/zoom cameras can reduce the number cameras necessary to watch an area. There are no panaceas. A PTZ camera programmed to pan across a parking lot will rarely (or never) be looking at what you need to see when trying to document an assault, vehicle break in, or other crime. As a cost justification, consider that a PTZ camera costs roughly the same as 2-3 fixed cameras. There may be some mega-pixel (extreme high-resolution) cameras that may be a better choice.

Generally, fixed position cameras are installed where they can provide an overview of what is going on, or to monitor a specific area of interest. Some of the common areas to deploy cameras are:

- Parking lots
- Pick-up / drop-off line & visitor reception
- Sports complexes / fields
- Gymnasiums
- Cafeterias and outside restrooms
- Playgrounds
- Vocational shop, computer and chemical laboratory areas

Many cameras now have the ability to capture audio. This can help to mitigate drug and gang activities but requires you to check your state laws and understand the legalities of audio recording.

Pan/Tilt/Zoom (PTZ) cameras are usually installed to compliment fixed cameras deployments. PTZ cameras can give unparalleled situational awareness when interfaced with other subsystems or directly operated by security personnel.

PTZ cameras, when interfaced with subsystems such as fire, intrusion, or panic systems, can automatically move to the area of interest and the video system can display the current onsite conditions almost instantly without the operator having operated the system. In this application, PTZ cameras usually reduce costs versus having dedicated cameras looking at specific assets or areas of interest.

During scheduled school events, PTZ cameras can help an operator to dispatch security personnel to resolve a potential problem or track down a person of interest. About 5 years ago, security personnel at a middle school near Houston Texas were able to capture the faces of vandals defacing property during an after-hour's event. Within minutes, the vandals were detained by security personnel dispatched with hardcopy prints of the vandals. With today's Smartphone technology, those images can easily be transmitted without having to wait to print out a hardcopy.

Adequate lighting is also a concern as you plan camera placement. Color cameras need more light than monochrome (aka Black & White) cameras. For this reason, most of today's cameras offer "day/night" functionality where the camera automatically changes to a monochrome mode. This feature allows cameras to be installed where lighting is marginal for color images and can sometimes save money by not having to add lighting. However, recognize that while you continue to see what is going on, you now lose the ability to distinguish the color of a suspicious vehicle or other aspects color helps to recognize.

Just a few years ago, thermal infrared cameras (which can operate with little or no light) were too expensive to consider for the average educational facility. Today, several manufacturers have produced very capable thermal infrared cameras that have an installed cost under $5000. Though still expensive, these cameras can be much less expensive than adding and maintaining lights and required power over time.

Recording & Chain of Custody

The obvious presence of cameras and a communicated policy of prosecution offer a significant deterrent to would be offenders. The majority of installed cameras are recorded as forensic or audit evidence. However, during an in progress incident, the ability to see what just happened minutes or seconds ago is extremely helpful. The above vandalism example relied on the ability to quickly playback video one frame at a time to get the best image of the vandals before they left the premises.

Stored video can also be leveraged as a force multiplier in operational concerns such as confirming appropriate staffing levels or evaluating fire drill procedures. These analyses, which depend on dedicated staff, can be augmented by leveraging stored video.

Facilities / engineering departments use recorded video to research pedestrian and vehicular traffic. Purchasing can confirm that deliveries were performed.

In each case, the actual operation of the video system should be performed by security personnel or a trained facility representative to protect the integrity of other video that may be introduced as evidence.

Recording Retention / Liability

Recording for educational facilities should be retained for at least four weeks, but longer durations are recommended. The decision of how long to record is a balance of cost, liability, and usefulness. Today, all recording is done digitally. The cost to store video continues to drop in relation to computer storage costs. Therefore, while cost is a concern, it is no longer a primary concern.

Liability in this case refers to how long you can reasonably be expected to store video and what is your risk should an incident occur and you cannot produce the video. This is a judgment call for your institution, and we recommend that you include your legal counsel in this decision. The existence of cameras creates an expectation of real-time monitoring and/or recording for the purpose of identification and be introduced as evidence. This exposure is not certain by any means, but we recommend that if you install a camera then you should record that camera.

Usefulness refers when the stored video becomes stale as related to the above-described operational benefits.

Chain of Custody

The introduction of video as evidence has become commonplace, but it does require that you have processes and policies to protect the integrity and acceptance of video evidence. The authors recommend that you obtain advice from your legal counsel to determine how best to reduce your liability and how your individual state laws affect your responsibilities (referred to as the chain of custody). In the event that video from your system needs to be presented as evidence, your security personnel will need to explain what the evidence is, where it came from, and where it went. To assist with this process, many video recording systems offer evidence integrity features such as digital fingerprinting, audit trail logging, and proprietary playback software for confirmation of these items.

Digital Video Motion Detection, Digital Video Analytics & Facial Recognition

Digital Video Motion Detection (DVM), Digital Video Analytics (DVA), and Facial Recognition are some add-on features that can be useful in your security technology plans.

As consultants, we routinely are asked about something seen on TV that someone needs for their system. It is important to recognize that almost none of the video wizardry demonstrated on the CBS television show CSI: Crime Scene Investigation is real.

Sorry, but the security industry has not progressed to the point of near instant facial recognition and discerning a license plate from four blurry pixels.

Many video recording and management systems and some cameras include built-in Digital Video Motion Detection functionality. The idea is straightforward; create an alarm when something moves into an area of the screen. The problem with this technology is that many of these systems cannot discern between shadows, and are just as likely to alarm on a dog versus a person. The best use for this type of technology is in areas where you control the lighting and have few opportunities for false alarms. This is usually adequate to detect someone moving in a hallway after-hours.

Digital video analytics takes this a step further and employs complex algorithms to ignore things like snow, rain, and shadows from clouds. Some DVA are able to do object classification and determine a human is, in fact, a human with a reasonable degree of accuracy.

Developments and cost reductions in the area of Digital Video Analytics (DVA) are allowing schools to add proactive detection components to their video surveillance systems. The "Object left behind" DVA offers a higher degree of accuracy and lower false alarms then previous video motion detection technologies. This means cameras can more accurately detect a person walking around a building after hours, while ignoring a wandering cat or dog. DVA can also detect a crowd forming or crowds dispersing quickly, which both have applications in an educational environment.

Today DVA is used sparingly due to a lack of understanding on how to use it without generating a large number of false alarms. Appropriate and dedicated camera placements are usually necessary to achieve a high degree of accuracy. We do recommend DVA but with appropriate planning and understanding.

Facial recognition is a technology that was hugely oversold when it first came into the market in the late 1990's/early 2000's. Like the wizardry of the CSI television program, Facial Recognition was promoted as able to identify faces in of a crowd. Today, the technology is useful but requires dedicated and sometimes specialized cameras. Though facial recognition uses cameras, it is not a video surveillance tool. Facial recognition is most effectively used for biometric verification of an access control credential.

Access Control & Visitor Management

Controlling access to your facility will significantly enhance Deterrence efforts in your security plan. In particular, your security plan should identify ways to reduce access control points and where possible locate those points near personnel who can monitor traffic.

Once your facility is controlling the entry and exit traffic patterns, then access control policies can be implemented along with efficient use of access control systems and security cameras to provide substantial a deterrent.

A physical security access, control system restricts entrance to one or more properties, buildings, or rooms to authorized persons. This feature can be used to electronically lock doors at predetermined times, reduce the issuance of keys, and make an audit trail readily available to the administration.

An access control system can act as a deterrent for internal theft and negligence because its presence alerts employees and visitors that your facility is serious regarding security. Most access control systems include the ability to track and log access to an area. Because employees know that they are being monitored, this has the effect of "keeping the honest... honest". Periodically making employees aware that the system lets you review access logs (audit trails) will keep employees from loaning their credentials to a student or other individual.

Access to a restricted area is typically facilitated by a person presenting a security credential (such as a card or key fob) to a reader. If they are authorized to use this area and it is during the authorized time, then the door will unlock. For additional security or alternate methods of authentication can be employed using, biometrics (hand geometry, fingerprints, facial measurements, or eye characteristics).

Likewise, students and visitors can be kept out of areas they are not supposed to access. For example, students should not be accessing cleaning supply rooms or high dollar equipment storage unattended. As well, visitors should not be able to gain entry and wander the halls.

Typically, authorized persons will include at least administration, security, and maintenance personnel. At the college and university level, an access control system can have associated benefits by including students (Dormitory access, library, labs, supplies, etc.). Elementary, middle, and high schools are still learning the benefits of enrolling students in access control systems as a universal ID.

Badging

Badging has enjoyed a lot of conversation and controversy concerning being able, at a glance, to tell who does or does not belong in any given school, university, or institution. Younger children tend to forget things like badges or even lunches, homework, and books. Badging everyone requires making many temporary badges in a primary and middle school environment on a daily basis, and thereby, increased administration headaches.

In primary schools, a more practical solution is to require visible badges on all adults. Faculty and staff members in your educational facility should have permanent badges that are always visible. Visitors should have temporary visitor badges thereby indicating that they are there with permission or are valid temporary employees and administration is aware of their presence.

Temporary badging solutions are available from a variety of sources and are inexpensive and easy to use. Supply costs are modest and worth the cost for security and reduced liability. From a policy perspective, training staff and students to ask for ID if not displayed, they can quickly identify a potential intruder or someone that does not belong in the school or has not followed the procedure to be properly identified.

Visitor Management

Visitor management is an extension of a well-designed access control plan and is intended to identify outsiders. Visitor management can include policies and procedures to make your employees accountable for their visitors. Additionally, if the building must be evacuated the system can be used to determine the presence and locations of visitors within the facility.

Digital visitor management systems can quickly verify visitors credentials against a national database of criminal offenders (Criminal Offender Records Information aka CORI), and provide an audit trail that is easy to create and access.

Visitor management systems may also include visitor-badging functions previously described as a feature of an access control system and relies on a controlled entry and traffic points.

Intrusion Detection

Intrusion Detection is often a separate system, but we recommend the operational inclusion of it as an integrated component of the access control system. If you buy a separate system, without the ability to interface or integrate, then you run the risk of performing a forklift upgrade in the future. (Forklift upgrade refers to the replacement of a large portion (or all) of a previously installed system.)

Intrusion detection typically encompasses the monitoring of doors, windows, overhead doors, air conditioning or duct systems, exterior vibration, and glass breaking. To accomplish this, these systems use several types of intrusion detectors, which are generally product agnostic for use with any intrusion detection manufacturer's panels. You are not required to use a detector of the same manufacturer with your system.

Each of these intrusion detection sensors is connected to alarm panels, which provide a centralized annunciation point. Intrusion detection systems can be configured with flashing lights and sirens intended to scare away intruders or to alert on-site staff. In some situations, it makes sense to install panic buttons in reception or the nurse's office. These systems can also be configured with a silent alarm to alert law enforcement.

Fire Alarm Systems

Fire alarm systems are required by law and heavily regulated as a separate life safety system. Most people would believe that if they have a fire alarm system then they are all set. However, this is not often the case due to ever changing code requirements, expired grandfathered systems, antiquated components, and temporary buildings with different/exempt regulations.

Schools are rarely given the amount of property protection that their importance dictates. Most schools are self-insured to a great degree. Losing a school is a disaster to the local community and the school district. We are forced to build schools at a much greater rate today than ever before and losing a school is going in the wrong direction and places a terrible strain on already overburdened resources. Economically it only makes sense that all schools be properly protected.

School fire alarm systems at the very least include smoke detectors and manual pull stations. Frequently these minimalist fire systems will only have a rudimentary alarm evacuation system (alarms that get your attention and signs to identify exits). Systems of this type offer no property protection and are typically not linked to other components of a master security/safety system. Systems that are more capable exist, which provide voice instructions and interface with systems for better situational awareness.

For example, by local fire code many fire alarm systems unlock the doors when a fire is detected. Burglars may use this to gain access to a building. If the fire system is interfaced with the video surveillance system, cameras near the alarms can automatically be displayed to dispatchers. Alternately, cameras can be used to identify when someone pulls the manual fire alarm because they under duress or pulling a prank.

School Bus Security

As discussed earlier in this book, parents expect their children to be safe from home to school, in school and on their bus ride home. It is incumbent on schools to be able to provide and plan for children's well-being from portal to portal and back again.

School transportation officials are learning that security and emergency planning on school buses is as important as security on the property and within the buildings.

In 2006, the National Association of State Directors of Pupil Transportation Services (NASDPTS) called for renewed security training for school bus drivers. At that time, the NASDPTS recommended school bus drivers receive training on personal safety, parental custody issues, crimes (vandalism, theft, and assault), hijackings, and possession of dangerous materials or weapons. The reasons for their concern are still present today.

These specific concerns can be mitigated through the use of security technologies such as mobile video recording systems and advanced video analytics such as license plate recognition. Today we see many school districts implementing access control cards for student riders.

We discuss some the technical challenges and options regarding school busses in the next chapter.

CHAPTER 5 - APPLYING SECURITY TECHNOLOGY

In earlier chapters, we have introduced security technology as a general overview. In this chapter, we examine the technology in greater detail to present important technical considerations and review desirable features and capabilities. In the next several chapters, we shift away from the technology and focus on specifying and implementing the security technology. If need additional references to expand your understanding of a particular technical area, we include resources in chapter 11 and offer details for self-help learning in Appendix A.

In order to determine what is required for your security technology plan, the operational capabilities of available technology should be well known. As with Chapter 4, we will start with Tier 1 systems (the tier closest to the user) and work our way through lower tiers. As we progress, you will be introduced to more specifics about the individual systems. We highlight some of the most common components for each system, and discuss some the more common integrations.

Experienced practitioners may have strong opinions when it comes to the selection of security technology. Many of these opinions are based on experience and personal preference. As with most applications of technology, there is always more than one way to solve a problem.

At publication, the technologies and features we address here are available in most mature manufacturer's security systems.

Tiered Approach Revisited

A primary concern in designing the security system is how it will handle all of the associated information produced by the various physical security systems. For example:

- Alarms from the Intrusion & Fire Systems
- Alerts from the Access Control System
- Video from the Video Surveillance System

Each of these individual system types has evolved separately and often present information differently. In chapter 4, we introduced the concept of a ***tiered*** approach to Physical Security Systems application. In review, this approach allows each purpose-built security system to operate independently, but allows the exchange of information between tiers.

The intention of this communication is to allow system information to be consolidated and presented by the Tier 1 system. This allows a standard user interface and the ability for the systems to present information more efficiently and more intelligently (For example, a door alarm can prompt a camera to be displayed).

Historically, this approach has been restricted by the willingness of manufacturers to develop interfaces with other manufacturers. With the introduction of Physical Security Information Management (PSIM) systems to the industry, we now have Tier 1 systems that interface to many more disparate systems and present information more intelligently.

Physical Security Information Management (PSIM) / Command and Control

Smaller educational facilities may feel their size does not warrant the addition of a PSIM. However, PSIM is a good place to continue our discussion because smaller, independent facilities are often brought into a campus wide plan.

As we discussed above, a well-designed, comprehensive security plan is not just about sensing and reporting alarms or recording and displaying video, it is about collecting and presenting all of the information in a coherent fashion so it can be effective used by security personnel.

Physical Security Information Management (PSIM) systems, previously known as Command and Control, allow disparate information from differing manufacturers' products to become interoperable on a standard platform or "user interface". Most PSIM manufacturers are system agnostic, therefore most manufacturers do not see them as a competitor and will be willing to cooperate and work with them.

One of the major advantages of utilizing a PSIM is to provide a "big picture" or "situational awareness" this serves to provide everyday benefit but can be extremely important in the most critical of emergency times. Most PSIM's can deliver additional benefit to the common operating picture with the ability to present a "live" campus system map. This map displays the systems in real-time, and the operational status of all the lower tier systems using icons to indicate the status of doors, other sensors, cameras and other system attributes.

Situational awareness is expanded when it incorporates other outside systems (non-physical security) monitoring (i.e. Fire, Mass Notification, SCADA, manual notification systems, HVAC, and other detection or notification systems).

Most PSIMs feature independent "modules" for each manufacturer's products (differing video systems, access control, etc.). The benefit of this modular architecture is that disparate systems or versions of systems can operate under a single PSIM interface. This can be very useful for campuses where independent systems have been installed in different buildings over the years as the budget allowed. In some cases, PSIMs can unify these differing building systems to a common operating picture without the penalty of upgrading all the sub systems.

Another important PSIM feature is the ability to employ business rules. Business rules can be simple cause-and-effect functions, or they can be applied across all of your subsystems, and address necessary employee response.

The term "business rules" in the security world has little to do with business. In this case, they are the processes and procedures as they pertain to your security systems and your security personnel.

PSIM business rules can cause automatic actions such as; "if a motion detector in the hallway generates an alarm, then show the associated camera on monitor X". The cause-and-effect benefit is operators do not have to waste time looking at empty hallways, or figuring out what camera to call up when an alarm comes in.

PSIM information and archive features can improve situational awareness in ways that other systems cannot. For example, the simple business rule example above can be expanded; by using the automatic call up, the PSIM could tell the operator if any other alarm (fire or intrusion) has occurred recently and show archived video on another monitor. With this additional information, security personnel can better determine if an isolated incident on one system is actually a pattern of connected events (such as someone testing the limits of your security).

PSIM's can prompt your security personnel to a specific response for a specific detected activity. Taking this example another step, the operator may be required to file a report using an onscreen electric form. In the case of a second alarm in the same area, the operator may be directed to dispatch security to that area.

The examples presented above, have been kept intentionally simple. In reality, PSIM notification rules and programmed responses can be developed for all types of emergencies, including the operator or administrator required actions, manual notations, and emergency action items. Until your security personnel have lived through a security or life safety incident, you cannot be sure how they may respond. Having a system that directs the operator to perform specific actions during an incident will keep them on the right track and deliver an appropriate response.

PSIM software requires computer servers, which are generally sourced as Commercial-Off-The-Shelf (COTS) product. With the importance of the tasks the PSIM system performs, we highly recommend that the servers include fault-tolerant components and employ a geographically separate hot standby architecture. PSIM's rely on LAN's (Local Area Networks) and WAN's (Wide Area Networks) to monitor and control Tier 2 subsystems such as Access Control and Video Surveillance. This allows them to be placed separate from the systems they are controlling, but requires a dependable infrastructure.

We have found the PSIM command and control environment works best in a dedicated security-staffing environment and a firm policy of response. However, PSIM also works well for consolidating disparate systems or implementing complex tiered integrations.

On the following pages, we will talk about the systems that can function both independently or as Tier 2 systems under a PSIM.

Video Management System Platforms

On the surface, many Video Management Systems (VMS) look the same. At the most basic level, video management software products route and/or record compressed video streams from cameras and encoders, and display video to monitors. These systems also provide camera and user administration and access rights. Most of today's Video Management Systems (VMS) display live video in a graphical user interface (GUI) or a standard Windows internet browser on an operators PC. Some systems offer video "decoders" to display video directly on dedicated monitors or "video walls"[1] without requiring a PC.

A close examination of available VMS reveals a wide variance of product features, functionality, and price. VMS manufacturers often times vary in maturity and market focus resulting in product differences in scalability, network management, fault tolerance, operating systems, software clients, and the availability of standard conventions and protocols (interfaces).

Video Management Systems require at a minimum: cameras, a PC platform with servers and storage equipment, and the manufacturer's software. Some manufacturer's designs are proprietary and require you to use their hardware, and/or software. Others manufacturer's feature open designed platforms, which provide the opportunity for easier system functionality additions.

---

[1] Video walls are often an independent subsystem, which allow camera video, cable/satellite TV, and other PC type information windows to dynamically share a "wall" covered in monitors. Video walls require dedicated hardware and software, which is integrated into some PSIM and video surveillance systems.

Maintenance and support of non-proprietary hardware is generally easier and does not require specialized or costly training or support contracts. Functions such as Increased storage capacity (external, SAN, etc.), software based firewalls, virus protection software, or intelligent video algorithms to name but a few can be operated in harmony with the video management software program.

We have found the broad range of VMS often necessitates a formal needs analysis and supporting trade study. After the needs are assessed, compatibility with legacy and future products (support of the third party SDK's and API's) must be thoroughly evaluated to insure that you are addressing the specific needs of your system requirements.

While you are determining the required functionality for your VMS, you will also want to identify if the manufacturers have the ability to integrate with your other systems such as access control, building management, industrial SCADA control, and business integrations such as Active Directory.

The term "business rules" in the security world has little to do with business. In this case, they are the processes and procedures as they pertain to your security systems and your security personnel.

Video Surveillance / Closed Circuit Television (CCTV) Components

Video Surveillance or the legacy term Closed-Circuit Television (CCTV) began with the use of analog cameras and display equipment that used one cable per video transmission. In CCTV systems, a cable is required for each camera to transmit its signal to a monitor or possibly a VCR. This is an over simplified description of CCTV, but it describes the basic premise of this technology.

Today video surveillance is making a shift from analog cameras (based on the original one-signal-per-cable design) to network (IP) cameras (where many cameras can transmit on network cable). There are still many analog cameras being installed because integrators are comfortable with their installation and troubleshooting. The cost difference between analog and IP systems is narrowing. However, the authors are hesitant to recommend analog cameras, as they will be obsolete in the near future.

As mentioned in Chapter 4, you may hear references to CCTV, Digital Video, Video Surveillance, Security Cameras, and the 'mystical' Digital Video Analytics. We would like to again point out that these terms are not interchangeable. Video surveillance is a rapidly changing segment of security that is much like computers, whatever you buy today will no longer be state of the art tomorrow. This rapid change and transitioning in the market (often referred to as "convergence") requires a well thought design concept to insure the long-term viability of your installation.

The key to extending the life your video system is to install a platform that supports "open protocols". This term means that the purchaser will not be limited to proprietary technical components from a single vendor. In theory, future incremental enhancements then will be cost effective and take advantage of new technologies without having to incur sweeping replacement costs for major system components.

Cameras

Cameras are the most basic and important element of video surveillance and predetermine operational characteristics of the entire system. The cameras you install predetermine certain system-wide operational characteristics such as:

- Fixed position or Pan/Tilt/Zoom (Can I remotely control the camera?)
- Standard resolution or megapixel (How wide of an area can I see or can I digitally zoom?)
- Built-in storage (Does video recording impact my network all the time?)
- Built-in video analytics (Do I get alarms from my camera that other systems can use?)
- Built-in processing power for future video analytics (Do I replace the camera or add hardware to use video analytics on a camera?)

Unlike consumer cameras, security cameras are built to operate on a full time 24/7/365 basis. High-resolution (Megapixel) cameras are becoming common in overview scenes and large area monitoring because they allow additional functionality. Recognize however, that higher resolution brings additional technical considerations and storage requirements.

Most recently, the cost for night vision and low light cameras have become economical enough to be used in commercial and educational applications to detect and capture images not previously available in older systems.

Motorized Pan-Tilt-Zoom (PTZ) cameras can be manually operated from a PC workstations virtually anywhere you have network connectivity. PTZ cameras can also be pre-programmed with rule sets to zoom in on detected movement and alarms from other systems.

Fixed position cameras in both the analog and digital IP configurations most often include auto iris lenses to optimize the picture quality. Today, there is a sweeping industry move where the auto***focus*** feature previously only found on PTZ cameras is making its way into fixed cameras. Having both Autoiris and Autofocus on fixed cameras helps to speed the installation and reduces maintenance adjustments by not having to open the camera housing.

Network cameras (also known as IP cameras) are replacing analog cameras because they can leverage existing network wiring and wireless networking technology to minimize installation costs. We talk about some of the concerns and advantages in more detail later in this chapter.

With the industry shift toward domed camera housings, outdoor cameras are now more aero dynamic and provide lower stress on the pan/tilt mechanism from wind resistance than the older conventional camera housings with exposed pan/tilt assemblies. Wind resistance can create maintenance issues when mechanisms are constantly operating to fight the environmental conditions. Not only are dome enclosures more aesthetically appealing they provide a better environment for all of the operational components as well as a more secure installation.

IP cameras use Internet Protocol to transmit video images and data control signals over a network connection. IP cameras are also commonly referred to as network cameras. When these cameras have additional functionality (video analytics or onboard storage), they can be categorized as an "edge device"[2].

IP cameras can be used for surveillance in the same manner as the more common (until recently) analog closed-circuit television cameras. A number of IP cameras are deployed together with a digital video recorder (DVR) or a network video recorder (NVR) to form a basic video system.

---

[2] "Edge device" in the security industry has a different meaning than the IT industry. In the security industry, "edge devices" are source devices (such as cameras and video encoders) that transmit physical security information to the headend, usually over a LAN/WAN.

Cost Advantages of IP Cameras:

- Providing reduced cost and enhanced functionality through the use of the IP networking equipment infrastructure.
- Typically provides a lower cabling cost (CAT5e or CAT6 instead of RG-59 coaxial cable).
- Lower space utilization in larger camera installations due to the camera switching and routing through the network infrastructure of firewalls, switches, and routers.
- Indoor camera power labor can be reduced using Power over Ethernet (POE), which allows power, video, and control over a single cable.

Other Functional Benefits of IP Cameras:

- Support for a variety of image resolutions including both standard analog CCTV resolutions and megapixel camera resolutions.
- Providing the capability for digital zoom of high-resolution megapixel images.
- Convergence onto new or existing IP cabling infrastructure control for PTZ (pan, tilt, zoom) cameras can travel over the same single network cable as video.
- Password control to prevent unauthorized personnel viewing images or altering camera configurations.
- Most IP Cameras will support different compression formats helping to relieve bandwidth and data storage demands.
- IP cameras allow encryption of camera controls and the audio/video information transmitted on the network.
- Some IP cameras have internal storage or processing to allow on-board video motion detection or other video analytics.
- Utilizing IP cameras provides benefits for integrating video surveillance with other systems and functions like access control, alarm systems, building management, etc.
- Remote configuration, network diagnostics, and maintenance reduce trips to components in the field.

The following is a short list of potential weaknesses of IP cameras in comparison to analog CCTV cameras. We mention these items as we feel that they are important considerations when planning and selecting the right mix of equipment for your environment and situation.

Possible disadvantages of IP cameras to consider in planning:

- Higher initial system cost if IP infrastructure not present or not robust enough to handle video load.
- Inconsistent video compression/transmission standards. IP cameras are available in many different models and use a different programming interface or may encode video differently. This means you cannot assume that IP cameras will be compatible with other IP video recording solutions.
- Technical barriers - Installation of IP Cameras require a different skill set to provide a concise group of network settings including IP address, DDNS, router settings, and port forwarding. This step requires network and IT structure knowledge not possessed by most users, without help from an IT professional.

Monitors

Video surveillance monitors come in an assortment of sizes and technical functions. Monitors can be applied differently as your monitoring layout and overall design strategy requires. Many Video Management Systems utilize one or more flat panel monitors of larger size to view multiple images. Often VMS, Access Control, or PSIM display additional system information, as well as video.

Multiple monitors maybe used, but you need to consider how your security personnel will use them. A good rule of thumb is that the more monitors security personnel are required to watch, the faster they develop monitor fatigue and are more likely to miss important events. A properly designed system can accommodate a minimal number of monitors but display all required information by using alarm point

monitoring, motion detection, and programmed video tours.

Monitors need to connect to some device to function. This will often be a dedicated computer, a network appliance (video decoder), or a "video wall" controller. Monitors can be used for viewing alarm notices, incident reporting tasks, and interactive maps. For this reason, expect that your final selection of monitor sizes and layouts will be ultimately determined by the size and complexity of all the systems being monitored.

Recording

Today's digital video recorders (DVR) and network video recorders (NVR) allow security system designers to use any combination of the previously discussed features of analog and IP security cameras.

DVR's are computers or purpose-built devices that use internal video capture hardware with specific channel capacities of 4, 8, 9, 16, 24, 32, and 64 (with each channel being a camera). These devices are closed ended with a predetermined storage size, thus requiring the need for additional devices if the system is expanded.

NVR's are designed to work with network edge devices (IP cameras and IP encoders) rather than directly connect with analog (non-IP) cameras. An NVR is a PC Server or specialized embedded device sometimes with video management software. Because they do not have to divide processing power and system resources with an internal capture card, NVRs usually will have a larger video channel capacity. NVRs with no local viewing can handle even greater video channel capacity because they do not have to interpret the video data they are managing.

There are some hybrid recorders that offer both DVR (video capture hardware) and NVR (IP only), but they typically have a reduced total capacity of IP and analog video streams.

Both DVR's and NVR's are easy concepts for IT directors and tech savvy consumers but "build your own" is something to be avoided. "Build your own" is the practice of adding video capture cards and software to regular PC's and building your own DVR or NVR. There are many problems associated with this practice that most security or IT departments should not handle long term. There are also associated liability concerns, as there is typically no manufacturers' product liability insurance and third party protection involved. In the long term, they are unmanageable and inadmissible as evidence in the event of legal action.

Recording systems vary greatly in scalability (as we touch on above). Evaluation of camera types (standard vs. megapixel), calculation of the quantity of camera types, and the anticipated system growth must be considered carefully in your purchasing decision. This is especially important to consider with fixed input DVR's or NVR's with resident storage. Many manufacturers provide calculators to predict storage requirements for their system based on your intended frame rate, number of cameras, camera type, and compression. Recognize that just because a manufacturer says they use a particular video compression standard does not mean their bandwidth usage and storage demands are the same from one manufacturer to the next, so use the manufacturers' calculator. Confirm the manufacturer's calculators match your expectations (some calculators are based upon imprecise math that frequently does not provide an accurate conclusion).

External storage is generally used on larger system platforms. In the largest of systems, network storage of video is the preferred method. Additional consideration should be given your final design and how it will interface with your infrastructure.

Methods of playback from systems vary as much as the devices themselves, consideration of this factor is important for the system growth, and how you intend on utilizing the system. In many smaller, underpowered systems (underpowered for price competitive reasons), you are bound to either a recording mode or a playback mode, not both at the same time. More robust systems provide proficient handling of larger quantities of cameras, and the ability to record and playback simultaneously. As you progress up the ladder to larger VMS systems with multiple workstations, functionality and versatility of recording, compression, archiving, and playback capabilities, generally increase.

When selecting the right recording device or platform additional thought should be given to the processing required for any analytics or additional functionality that the system will perform now or in the future. Keep in mind that processing power is what determines the success and speed of your system operation. As we have noted previously, this is an area where it helps to educate yourself or solicit expert help in your system selections.

Digital Video Analytics

Digital Video Analytics is not the "magic" we see on TV or in the movies, in fact, there are specific applications where video analytics (of a certain type) work, but more where they do not. When deciding to utilize a specific video analytic it is important to evaluate and consider what the intended result is. Many manufacturers and products of the past touted inflated claims of performance, which in turn yielded poor or ineffective results, which translates to no alarm detection, or more often, false alarms.

Various video analytic products exist in the marketplace, some of which are placed within the operational software (within the NVR, DVR, or VMS platform) system and some of which are placed at the "edge" in the camera. While no methodology is distinctly better, some analytic software performs better than others do, and some combinations are more cost effective than others are. One of the lessons learned by designers of systems utilizing digital video analytics is to pick a system based upon the required results and use dedicated cameras selected and placed specifically for the application.

Typically Found Analytic Functions

Motion detection

Motion detection is considered the most basic type of video analytic. Video motion detection processes each pixel in a video frame over the period examined. If there is a significant change in brightness, contrast, or a change in pixel intensity an alarm is created. Some systems address the speed in which a change takes place or where in the view motion detection occurs to manage false alarms.

Motion detection is utilized to create alarms, which alert security staff to unauthorized activity. These methods use motion detection for many public safety applications by creating alarm areas that are not to be entered. In post-incident review of recordings, specific areas within a scene can be searched automatically utilizing the motion detection function to find and view specific occurrences.

Object Classification and Tracking

Object classification is achieved by comparing camera video with known shape models as they move in a given scene. This video analytic method attempts to recognize pre-described objects such as a human, motor vehicle, and small animal. Once recognized this determination can be used with other analytics like video motion detection.

Object tracking video analytics can also be coupled with PTZ (Pan/Tilt/Zoom) cameras to track detected objects as they move through a scene. This method is often a function of motion detection combined with the tracking functionality addressing pan/tilt/zoom cameras for tracking to work well.

Congestion Detection (Crowd Forming/Crowd Dispersing)

Congestion detection is similar to motion detection but represents a more complex setup to work correctly and provide usable detection. This type of video analytic monitors the density of humans or cars in a given space. An alarm is triggered when this density exceeds a set threshold in the analytic software. Security personnel are notified that a potential situation could exist. These alarms can be triggered if a crowd starts to form or if normally constant foot traffic suddenly disperses.

Object Removed

Object removed video analytics, as its name indicates, can be used to create an alarm if an object that remains in a stationary position in view has been removed. Object removed can be used in libraries, offices, art displays, and a host of other applications to protect items from being removed or when you wish to insure an object is left undisturbed. Camera view is extremely important when utilizing this type of analytic.

Object Left Behind

The Object Left Behind analytic looks for objects in the field of view such as suitcases and bags left unattended for a programmed amount of time. This system function is useful for monitoring public areas where an unattended item may not be discovered by security personnel for some time (often used for bomb threat detection).

This analytic can also be used for monitoring time-limited parking spaces, school drop-off areas, or standing zones. Personnel can be alerted that a parking offence is occurring or has occurred.

It is important to note that this type of detection for small objects (suitcase, briefcase, bag, or box) requires camera views of specific areas that are not obstructed from view by people or other objects entering and leaving the field of view.

Wrong way detection and object counting

Wrong way detection analytic detects objects moving in the wrong direction with respect to a permitted flow of traffic direction. This analytic is especially useful on a school campus to identify a potential problem during high traffic entry and exit times (sporting events, registration, assemblies, etc). The analytic can also be applied to your system to identify vehicle traffic travelling the wrong direction on a one-way road to create an alert or alarm. Some analytics manufacturer's platforms also provide counting of people or vehicles through an area allowing a total count in a direction.

License plate recognition

License Plate Recognition (LPR) software is a specialized video analytic type where video streams from cameras looking at vehicular traffic are analyzed, to locate and isolate the license plate in the video frame.

The LPR software is then utilized to compare the license plate identity to a database of plate numbers in an approval or alert/alarm fashion. The LPR analytics database contains user rules based actions to be used for security staff.

License plate recognition is the most specialized of the analytics and generally is a specialty platform or product to produce consistent results. LPR may also require the use of specialized lighting or cameras to produce the intended result.

Facial recognition

Facial recognition is perhaps one of the most misunderstood and misapplied video analytic technologies today. As of this writing, no proven video analytic for public use has an acceptable success rate to identify a face in a crowd or a known face from a crowd.

Most if not all successful applications of facial recognition require a well-lit, unobstructed view of a single person's face presented as a validation to a credential (access control card). While we are interested in this technology, we have been hard-pressed to find a viable application in a school or campus environment given the cost and processing delay time.

Limitations and Conclusion for the use of Video Analytics

Though they have been around for some time, video analytics are still not well understood by all for the correct use and implementation. New advances and product offerings are available each year, but should be evaluated carefully.

One of the primary interests for using video analytics is to intelligently monitor and process multiple cameras. However, video analytics has not progressed to the point where it can replace human decision-making or make judgment calls. There are many common current misconceptions regarding video analytics abilities to detect and analyze information.

The most advantageous video analytic that we have seen provides users with video information that is usable by triggering only incident-related alarms that need to be viewed, as opposed to watching long periods of video without any substantive change or activity that needs response.

Some video analytics can be coupled with actions that can be automatically started to create warnings or deterrents to illegal, dangerous, or unauthorized activity. A system can create audible warnings, turn on lighting, raise or close barrier gates, lock doors, and a host of other functions based on the system rules.

Video analytics can be cost-saving when used prudently. Some analytics can perform routine tasks that often become mundane to security personnel. By supplementing your monitoring location and combining portable devices and cameras systems with analytics your system can be used as a "force multiplier" for limited staffing during events or emergencies.

Large-scale video installations have many areas with little or no activity being monitored for long periods can benefit greatly by implementing video analytics. Video storage requirements, network transmission (bandwidth requirements) can be reduced by processing only the required video with analytics.

In every video analytics deployment, camera placement and properties have a profound effect on performance. Proper expectations and adequate design are important to insure a successful deployment.

Mass Notification Systems

Mass Notification type events can vary from the severe "a bomb or terrorist attack", to the informational such as closings due to weather "a snow day". The intent of Mass Notification Systems is as the name would imply "to notify the masses" in the case of an event.

As authors in this subject matter, we would be remiss not to caution the reader that Clery Act considerations and changes within NFPA, set expectations and requirements regarding these systems. We advise that these items be given the consideration they deserve. Laws, regulations, and codes have begun to change regarding fire and notification systems. When designing these systems NFPA requirements and recommendations should be followed closely.

In some locales, requirements utilize NFPA as a guide and add requirements to make the requirements more stringent. As you consider these systems, please keep in mind that while State and ultimately local jurisdiction presides, any deviation that is less than the NFPA standard opens the school or institution to potential litigious pursuits.

Many voice-based systems come preprogrammed with English, Spanish, or French messages, and often are available to use customized message(s) recorded by the customer. The power of the human voice added to the typical bell, siren, and flashing lights makes a system highly effective. However, too often systems are installed by the lowest bidder and are less than adequate making annunciation of voice completely unintelligible.

Careful consideration is needed, to ensure your message is effectively heard, the messaging is clear and "ominous or foreboding" enough. Verification and additional notification are also appropriate in larger settings, by this we suggest the supplemental or information monitoring by the PSIM system.

Many of the failing points in system installations are:

- Cabling that is not appropriate for the distance and number of speakers.
- Incorrectly sized speakers, or not enough speakers for the area.
- Indoor speakers installed outside.
- Underpowered or low sound quality amplifiers.
- No redundancy or overlap in the event of malfunction

Mass Notification systems should include or consider the requirements unique to your setting as follows:

At a minimum:

- Public address system
- Posted notices
- Emergency call-in hotline
- SMS/Text messaging
- Website/portal
- Outdoor Public Address System/Warning Sirens
- Voicemail notification system

Recommended:

- Digital signage
- Desktop alerting in each classroom and throughout the campus/school
- Voice enabled programmable fire alarm
- Mobile loud speaker system(s)
- Blue light towers
- Staff radio system with alerting
- Outdoor emergency phones
- Utilization of Social media, blogs, etc.

This list is not all encompassing, but a starting point to identify and evoke the discussion regarding the responsibilities and the tools available in your unique circumstance to protect the students, faculty, and staff of the institution.

Automated Phone and Text Messaging

Messaging systems are often routinely used for special town/city events requesting participation, weather related/snow day changes in schedules, or for other public service announcements. When the system is used as part of a total mass notification system, other automated or "hazard" information can be rapidly dispersed amongst a population.

Phone and Text messaging systems are primarily computer based today. While these systems can be used stand-alone, they are most effective when utilized as a component of a complete Mass Notification System. The intent of the messaging system or as a component of Mass Notification is to deliver a large volume of messages with the same content to many.

For this reason, consideration should be given to the capacity of how many messages can be sent out versus how long. If it takes an hour to get the message out regarding a life safety event, then your message cannot be considered timely. Conversely, public service messages can take longer without dire consequences. It is also extremely important to pick a highly qualified contractor with experience in the installation, federal and state regulation, and requirements.

Access Control, Badging and Visitor Management

As we state above in chapter 4, the purpose of an access control system is to limit access and create a record of accountability. Employees can only access areas where they are authorized, and during the hours they are authorized. Likewise, students and visitors can be kept out of areas they are not supposed to access. When access control is paired with effective visitor management, a complete accounting of the population is available.

Access Control and Badging

Access Control is significantly different from a simple burglar alarm systems because Access Control allows or denies entry in addition to monitoring breaches. Access control systems also act as a deterrent for internal theft and negligence. Most all access control systems on the market today include the capability to log individuals that accessed an area via unique user types (ID's) and a complete and ongoing record of system activity (archive). Because employees know that they are being monitored, this has the effect of keeping the honest… honest and employees are less apt to loan their credentials (linked to them) to a student or other individual.

Commonly, access control systems will allow the supplemental tie in and monitoring of multiple systems, i.e. fire, perimeter alarms, interior alarms, video system motion detection, and other alarms. It is very important to gather all of the disparate information and utilize it fully in a redundant fashion, as long as it remains well coordinated and usable.

The use of access control, in most cases, requires that the access control system be connected to the fire alarm system so that in the event of a fire alarm the emergency egress doors are automatically unlocked. This also requires the potential use of code-approved request to exit devices be employed in the systems.

The most basic component of security access systems is a keypad or card reader or some combination of both located at entry points you wish to secure. The design of the access system needs to be given careful consideration and planning with regard to use in order to achieve a successful deployment.

The keypad utilized within an access control system can sometimes have beneficial effects in that no credential is required to access the facility. In the case of educational institutions, we do not recommend this path. Credentialing is very important and highly beneficial (we discuss the benefits in the next section further).

Card readers (credential readers) utilizing magnetic stripes, RFID technology, and traditional weigand swipe technology are the most prevalent. RFID technologies include Prox (short for proximity) and smart cards, which are similar to Prox but house additional data other than just the user's ID. As you evaluate these choices, you may be tempted to tie your cafeteria or other card based purchasing program (sometimes referred to the One Card system). Our experience is that combining the security and commerce is generally problematic.

With the cost reduction of the RFID technology, it is of perhaps the greatest benefit from a field device standpoint. RFID readers are small, contain no exposed components, and are low maintenance. Other methods of verification that are available include Biometric/hand/finger scanning and other devices (ex. Retinal scanners). Regardless of the initial sensing/access granting, each device should allow access to separate and specific zones, areas, or rooms so that your users can be grouped together with their access areas.

The door access devices are wired to control panels and from there to a computer system (Server or other manufacturer specific PC) to run software. These panels are generally proprietary to the manufacturer. All access control manufacturers' software is proprietary, and therefore this decision is of paramount importance. Examine all of the performance details of the equipment to determine if, and how they meet your institutional requirements now, and for the future. It is also important to ask for referenceable installations of similar size and scope to understand that the manufacturer and not just the integrator has experience in your type and size of installation.

Some access control systems allow the interconnection of components or panels over non-proprietary wiring, or more specifically over your existing LAN network. It is important to understand how the system functions, the traffic that is created on the network, and the level of isolation that is required from generalized network traffic prior to proceeding with this type of installation. We have seen IT Directors accidentally disable an access control system because it was doing something he was not expecting. The network based access control system can provide significant benefits provided the controls and network capabilities exist to support the system.

Another advantage of network-based systems is that many of them include system wide diagnostics provisions. These diagnostics features will pay for themselves repeatedly by reducing system operational errors, quickly identifying problems, reducing repair time and cost, and ease of securing a large facility.

As previously mentioned, most access control system components can be interfaced to other systems such as time and attendance systems, visitor systems, and administration systems to obtain multiple functions from the single investment (this integration is much easier and economical when all of the systems can be tied through your network). This can be helpful but recognize that if your integrated access control system fails, you will lose your larger operational picture unless a PSIM is included in your design.

Visitor Management

Visitor management in school is an important extension of a well-designed access control plan, and is intended to let the right people in and keep the wrong people out. Software based visitor management systems can quickly verify visitor's credentials against a blacklist or a whitelist (CORI, and validate parental rights of custody are just two efficiencies created), and provide an audit trail that is easy to create and access. Beyond the obvious security improvement, visitor management makes your entire staff accountable for visitors while on site.

Ties between the access control system and the visitor management system can aid in accountability if the building must be evacuated. Most robust Visitor Management systems can be used to determine the presence of visitors within the facility or provide a "muster list" to account for all evacuated guests.

Visitor management sometimes operates within the access control system in some manufacturer offerings; however, we recommend stand-alone software so "best of breed" integration can be achieved.

Several visitor management systems have the ability to integrate with third party access control systems to share access control readers and door controls. This integration provides an access related audit trail specific to each visitor. By leveraging this type of integration, it will save you from duplicating hardware or installing disparate hardware.

The efficiencies of Visitor Management are tied to the software offering and how robustly it can be programmed (rule sets) as well as the necessary ties to products in the physical security environment (access control & PSIM). We recommend that much like other products that are software and integration intensive, that educating yourself and/or seeking professional help will deliver the best results.

Fire & Intrusion Central Monitoring

As we described above and in chapter 4, there are several ways to approach the monitoring of your security and fire alarms systems. We will explore several options below. Please keep in mind, as we have stated earlier, the best system designs incorporate layered redundancy, which in the case of fire and intrusion systems lends itself to supplemental monitoring by a hierarchal system.

All state of the art Fire and Intrusion systems contain digital communicators capable of transmitting to a central station or monitoring center in a variety of formats. However, you will have to make sure your systems support a format that is compatible with whatever central station or monitoring method, you elect to use. We talk further about the distinct components of a fire system later in this chapter.

In the conventional configurations, digital receivers are assigned to dedicated telephone lines and integrated with the central station software. Field security panels send alarms, open & closing signals, supervisory signals, test signals, and periodic inspections of all systems to central station receivers. The central station software deciphers this information and displays it on computer terminals for action by the central station operator.

There is a growing segment of the security systems equipment, which can be centrally monitored over the Internet. Only specific equipment has been listed or approved to date. However, Internet Monitoring is becoming more wide spread in its use. Security has a responsibility to use the time-tested methodology.

User Owned Central Station or Monitoring Center

Owning, maintaining, and manning your own monitoring facility is an expensive proposition that should not be taken lightly. Central

Stations should be UL approved facilities with a host of specific requirements. To maintain this approval, facilities are subject to annual inspection fees from the Underwriter's Laboratories. The facility will also require specialized equipment such as dual hot redundant PC platforms and equally specialized power features such as UPS, battery backup and emergency generation. In many states, your central station will require a license and reoccurring operator training for the individuals staffing it. Lastly, the added liability may require additional insurance coverage and premiums.

There will be ongoing maintenance costs associated with the licensing, technical support and software upgrades, which also need to be taken into account.

There is a significant expense to operating your own central station. It is only reasonable to consider your own central station if you have your own police, fire, and security forces deployed 24 hours a day, seven days a week, and 365 days a year and are willing to hire fulltime dispatchers.

Independent Third Party Central Station

The third party central station will usually be UL approved thereby providing all of the assurances of outside-unbiased supervision and having all of the electrical, physical, ventilation, and software requirements met. Third party central stations are in the business to provide this type of monitoring, so they assume all the burdens of insurance, training, and any licensing requirements.

This solution usually is best for the school, university, or institution that does not have a full time police or security officer program to mitigate all of the associated costs.

It has the most flexibility with the least associated risk and it provides some sharing of the risk and liability.

Independent Central Stations generally do not sell or install intrusion or fire systems. To be competitive, central stations will usually support a majority of the alarm communication standards. This allows them to monitor many accounts spread the overhead cost over the entire monitored customer base. This type of service provider, you can select options and be billed for central station alarm monitoring services in an all-encompassing, or al-a-carte fashion.

Security Supplier Owned Central Station

In some instances where a school district, university, or institution contracts their security services to a security systems supplier, including monitoring, it is the responsibility of the security systems supplier to assume all of the previously discussed burdens of cost, insurance, training, licensing, and UL approval.

It is your responsibility to visit and review their central station insuring that they are, in fact, doing all of the things required under your agreement with them. In many respects, this method while giving your institution a single point of responsibility it also provides for a single point of failure.

Outages can occur from weather, construction, or an attempt to compromise the ability of the security system to communicate with the central station. Attacks on telephone lines are becoming more common during acts of burglary and arson.

Whichever monitoring solution you ultimately adopt, we strongly recommend that you do not rely solely on unsupervised telephone lines as your sole source of communications with the central station. We highly recommend that you employ backup solutions involving cellular communication back up, telemetry, long-range radio, or a network-deployed application to facilitate the delivery of emergency signals during telephone outages. You can access a list of these suppliers using the link provided in appendix A.

Intrusion Detection Systems

Intrusion Detection is an integral part of your physical security technology. Intrusion detection can be a standalone Intrusion Detection System (IDS), but is often part of an Access Control System. The descriptions below apply to both types of applications of the technology.

IDS sensors (described below) are connected using dedicated wiring. This wiring is often protected in metal conduit to prevent tampering or rendering a device inoperable by simply cutting the wire. In many cases, these sensors will be "supervised" to detect tampering or make the sensors inoperable.

Rather than run wire from every device back to a central point, IDSs usually have control panels located throughout the building in secure locations. As with the device wiring, IDS panels should also be monitored for tampering.

These panels serve as aggregation points for intrusion detection sensors and transmit alarm conditions to alarm monitoring workstations, annunciation devices, or central monitoring station. IDS panels should be connected to emergency power with battery backup to ensure the panels remains continuously operable.

When designing an intrusion detection system, it is important to use overlapping layers of sensors that support each other if one method fails. However, a very large school, college, or university could be expensive to protect if we tried to apply protection to every door and window individually.

Understand that when you diminish the amount of protection in any circumstance you increase the risk; we are not recommending that we become penny wise and pound-foolish, just understand the risk/expense equation. A qualified professional risk assessment that understands your specific needs can add valuable insight to your decision making process.

Balanced Magnetic Contacts

Balanced contacts are used on all types of doors and windows to detect when opened or closed. Not all magnetic contacts are created equal. There are high security balanced and biased contacts, but these are normally reserved for government or financial institution use. There are several operational characteristics worth noting when specifying magnetic contacts; the gap should be 1" or less and magnetic contact design should be selected for specific application.

These devices are often used for monitoring doors and windows of building exteriors and specific interior spaces. Some of these interior spaces include computer labs, bathrooms, locker rooms, auditoriums, kitchens, labs, libraries, band instrument rooms, and storage closets.

Motion Detection

The use of long-range combination passive infrared (PIR) and microwave detectors in corridors and hallways is becoming commonplace. These represent two different technologies, which can be applied to detect motion in different areas. PIR measures infrared (IR) light radiating from objects in its field of view. While very accurate, these devices can create false alarms from items that change temperature such as heating radiators, and electronics. PIR should also be placed so they do not "look out" windows.

Microwave detectors send out pulses of ultrasonic waves and measure the reflection off a moving object to detect motion. The technology works similar to a police radar gun. These devices are also very accurate but they can be bothered by electrical "noise".

Some motion detectors use a combination of both PIR and microware technologies; when the PIR sensor is tripped, it activates a microwave sensor to confirm the motion detection.

When laying out motion detectors on a scaled plan, always indicate the protection pattern. This information can be found packaged with each detector and on specification sheets. For example, a passive infrared intrusion detector (PIR) could indicate coverage of 50′ x 50′ with a drawing indicating the degrees in a fan configuration emanating from the device itself.

Any room that merits a door contact should also be considered for at least one motion detector. Other areas you may consider covering with motion detectors include vending machine and safes or vaults.

Other Alarm System Features

The security system can be used for a host of other supervisory tasks capable of alerting school personnel of potentially damaging events. Monitoring food storage temperatures in coolers and freezers, monitoring power outages and generator failure to start, HVAC Systems and in colder climates, boilers used for heating is a frequent application.

The security system can also be used to monitor the building fire alarm system (provided it meets the UL Laboratories listing).

In addition to the protection provided by the security systems, a well-documented deterrent is signage deployed throughout the facility identifying that these premises are monitored and protected.

Fire Alarm Systems

Fire alarm manufacturers are regulated and scrutinized by Underwriters Laboratories, National Fire Protection Association, Factory Mutual, insurance risk carriers, federal and state agencies, ADA, and many others. They are tested routinely in order to insure reliability. If they are so reliable, then what is the concern?

For fire alarm systems to be of any value, they must be designed, installed, and maintained correctly. If you already have a fire alarm system, the following information can help you to understand where you might lower some of your maintenance costs and how to integrate your fire system into your overall physical security plan.

Conventional Versus Addressable

Fire alarm architecture is described as conventional or addressable. Your choice of architecture determines the devices you can attach and some of your system's core functionality.

In both cases, there are two functional sides of a Fire Alarm System;

Detection and Annunciation. The ***detection*** side of a fire alarm system is comprised of detectors and pull stations that are connected to a fire alarm panel. Similarly, the ***annunciation*** (or alarming) side of the system is comprised of the same fire alarm panel and is connected to strobe lights and alarm sounders.

In a ***conventional*** fire alarm system, each side (detection versus annunciation) is typically wired separately. Depending on the fire alarm panel, a certain number of devices can be connected in parallel and each string of devices is considered a ***Zone***. Depending on local code and the fire alarm panel design, these zones may refer to an area of the building such as the "south hallway classrooms" or the "auditorium".

Conventional system components are often used because they are simple to design. In small to medium installations, conventional systems can also cost the least to install. However, the simplicity of conventional systems includes some drawbacks. For example, when a conventional detection device is activated, the system cannot determine which specific device is in alarm. Conventional systems can only detect that a device in a particular zone is in alarm. If your zones are large areas, this could be any one (or more) of dozens of devices. This makes debugging and maintenance more difficult and translates into higher maintenance costs during the life of the system (which should be factored into your total cost of ownership which we describe in chapter 8).

Addressable fire alarm systems have devices that are individually addressable, meaning that they are uniquely identified by the system. Hundreds of devices can be monitored through a single cable, but each device has its own address (digital ID). Instead of being associated with a zone by the physical wire the device is attached to (as in a conventional system), addressable fire alarm devices are assigned to zones through programming.

Addressable systems provide the responding fire department with information, allowing them to go right to the area of the alarm and investigate the potential fire or problem. Addressable fire alarm devices often have the ability to indicate a 'trouble' state, which makes them easier to maintain over conventional systems because you can tell exactly which device needs service.

The added flexibility of addressable systems sometimes comes with a slightly higher component cost and initial installation cost, but maintenance costs are generally lower over the life of the system.

Some hybrid "multiplex" systems can mix conventional and addressable system components. However, most system designers will opt for one type of system or the other.

Panel-to-Panel Networking and Proprietary Systems

In large facilities and campuses, you can expect to see multiple fire alarm panels (we recommend each building be separate from one another for reasons of fault tolerance). There are quasi-standards that exist to allow panels to be networked to each other (for example BACnet), but they are not widely used.

Device to panel wiring patterns, power requirements and communications protocols (in addressable systems) can also be proprietary between manufacturers. The reason for these differences can be as simple as the manufacturer believes their configuration is better in some way, or it can be part of a larger marketing/technical support plan causing you to buy all upgrades / additions from a smaller pool of installers.

Whatever the case, pay attention to what you have already installed or are researching for future consideration. Expect that once you start with one vendor you will need to stay with that vendor's panels, at least within the same building.

Fire Alarm Components

Smoke and heat detectors are your primary fire detection devices and should be installed according to code and located throughout the structure. Manual pull stations should also be located according to code and protected with covers that sound an audible alarm if the manual pull station cover is opened. This prevents false alarms and class disruption.

It is also important to remember all horn/strobes also must be specified and located in compliance with the fire code, Authority Having Jurisdiction (AHJ), and Americans with Disabilities Act (ADA). (See appendix)

There are many types of fire detection devices on the market today and most do a great job for their intended area of specialization. There are many mainstream and specialized detectors available; we recommend that you follow code requirements when deciding which of them you need.

Fire Alarm Monitoring & Integrations

In almost every state in the United States, the International Building Code requires fire alarm and sprinkler systems to be monitored by an approved supervising station.

Supplemental notification is allowed and with the advent of mass notification systems, fire systems can reside under the mass notification system according to NFPA rules. In any event, it is prudent to have supplementation notification by your access control system to ensure your on site staff has a clear picture of what is going on.

Building codes may require also integration with your access control system to unlock & open doors if your fire alarm system detects a fire.

Head-end / Critical Component Placement

Your physical security systems will have field devices that report to a ***Head-end***. The Head-end is where the software is installed to provide rules based decision-making, user management, and logging. Many of these systems contain intermediary components, like access control panels and DVRs, which are self-contained and continue to work if severed from the Head-end. You will need to locate both the head-end equipment and these intermediary components where they can be connected to power, network infrastructure, or in some cases dedicated wiring.

Most security equipment will need to be installed in 19-inch wide equipment racks or mounted on a wall. In the case of rack-mounted equipment, you will need to find space in existing racks or find space for new racks and purchase/install them as part of your project.

Do not allow this equipment to be placed under the desk as a footrest or in closets with no ventilation! This is important equipment, and should be installed with the same attention and forethought as your IT equipment.

In every case, you will need to make sure that there is enough ventilation to keep the equipment cool. Access control panels and fire panels are less restrictive than DVRs and NVRs, but all of these critical components have environmental considerations and ratings. When placing a component in a space that does not have dedicated environmental controls, we recommend installing temperature and humidity sensors to warn when the space is outside ideal ranges.

Care needs to be taken to ensure that critical security components are placed where they cannot be subverted or illicitly accessed. Site surveys often reveal DVRs in unlocked cabinets, on people's desks and under people's desks. The value of the equipment and its proper function need to account for the problem of dust and overheating. Owners that overlook the point that these units have security video that may be needed as evidence someday place themselves and their institution at risk for litigious action.

Securing cabling at the head-end is also a concern. It is a good practice to enclose entrance wiring to cabinets within conduit until it is not exposed or is above a ceiling and out of easy reach.

Power

Power is a critical consideration especially when you intend to use the security system to provide supplemental monitoring of the fire alarm system or other life safety systems. Physical Security and Life Safety systems require clean, uninterrupted power to ensure system stability. System components can be damaged by power fluctuations, just like your PC.

Power standby requirements will be different for each system and must be in accordance with the applicable Federal, State, and Local codes as well as manufacturer requirements for their system. Power requirements are dictated by the number and variety of devices attached to the control panel (s) and used in the system. The calculation and accounting of all system devices must be done in order to provide load specifications for the system backup and load draw from primary power. The individual device specification will generally indicate how much power is consumed normally versus peak consumption. Your calculations should consider simultaneous inrush (when all devices are switched on simultaneously).

The Security and Life Safety system power should be made available from a dedicated circuit breaker, identified in the electrical panel with a screw on protector eliminating accidental actuation or shut off. Security system power should always be in an area inaccessible to the public and have a door contact on the area entrance to prevent unauthorized entry and possible compromise.

Lastly, redundant power is not enough to support today's computer based systems. Backup generators and some switching systems have an unavoidable delay. Computer based systems can be damaged during this transition and operation will suffer from the forced restart. These types of components should be additionally protected by battery backup with ample time for failures of both primary and secondary power.

School Bus Security Technology

Technology has advanced rapidly to the point where it is economically feasible to track and provide two-way communication and private or public addressing to each school bus. This technology will handle an array of potential hazards that may be faced during the day while a bus is in transit. Video recording of events helps to deter and defend against bullying, assaults, and accidents.

In addition to the information that can be transmitted and gathered from the buses, this information can be utilized by state and local law enforcement even in real time from a mobile platform. Hand-off of video and data from buses can be efficiently handled and coordinated to work in harmony with school security and notification.

While our book is meant to focus on Electronic School Security, buses provide far greater ties to our communities than is often considered. When buses are well equipped, managed and prepared through planning and operation they can be a valuable asset to any community (New Orleans and the post Katrina incident should be enough to convince the importance of this issue).

Some important considerations that exist when presenting and detailing the need for the technology inclusion within your bus fleet are as follows:

- Is the technology designed for a mobile/transportation environment
- Hardened equipment with power backup
- Recording mechanisms that are hardened or solid state
- Protected from extremes of heat and cold
- Is the wireless technology incorporated in the hardware
- How are updates delivered to the hardware and software (directly or wirelessly)
- How does the system report faults
- Are the cameras designed for rigors of environment
- What other peripheral devices (lights, panic switches, geolocation) can be added to the system

Physical security technology on school buses and public transportation is currently a specialty niche serviced primarily by companies specializing in transportation. These systems sometimes need to interface with other physical security systems, so we felt it was important to give you this brief overview.

CHAPTER 6 - IT AND THE PATH FORWARD

Physical security technology, including video and access control, is critical to a school's efficient security operation and general well-being. Security has historically operated separately from other business functions, an arrangement that has supported its distinctly focused mission. Many of the core IT requirements for dependable, fail-safe operation are similar in many ways to Security and Life Safety. For this and other reasons we will discuss, it makes sense to want to use IT networks for security infrastructure. It also makes sense that the IT department routinely provides and maintains proven reliable PC servers that could service security system platforms. The truth of the matter is, due to some distinct differences in understanding nuances of requirements convergence has proven slow to translate from one industry to another.

The security industry has a demonstrated lag as compared to other industries. However, there is a discernible shift in the video industry, which is aiding the overall IT convergence. From 1980's into the early 2000's, the video industry has been dominated by large "full line" manufacturers who provide one stop shopping. Literally hundreds of manufacturers surfaced and disappeared during this time. A handful of the remaining companies have become leaders in certain specialty areas. Today, there are manufacturers who specialize in *only* IP cameras or *only* video management software. These and similar products are outpacing the "full line" manufacturers who continue to struggle to balance engineering costs over a broad range of products.

To be fair "full line" manufacturers are not the nemesis of convergence and the slow translation cannot be attributed to just one hurdle. Around, 2007-2008, a well-known network hardware manufacturer entered the security market by acquiring an emerging video surveillance manufacturer.

The security industry took notice of this new entry with both excitement and dread. Many manufacturers were concerned because IT technology was commonplace and well defined within most all organizations. Additionally, some were concerned because this IT leader has deep pockets and an intimate understanding of IT that many security industry manufacturers lack.

Now many years later this IT leader's video surveillance products are not approaching the top 10 of any security industry evaluation, their products are still not ONVIF compliant (currently one of the most accepted security video standards), and they are not open standards compliant. Previous successes in IT, deep pockets, and the desire to become a formidable player have not translated to success in the physical security marketplace.

For these reasons and other lessons learned, we recommend you disregard most of "convergence" marketing messages. Today, we recommend that you consider security technology as dependent on IT infrastructure, and give them both wholly separate consideration and resources with differing and distinct needs of their own. The rest of this chapter will give you some concepts to consider regarding IT infrastructure.

Security Technology Impact on IT

Today's more efficient LAN/WAN infrastructures are a natural alternative to the security industry's historic use of dedicated, proprietary infrastructure. Many school campuses have networks already connecting all the buildings on campus. This "primary" network (used for email, internet, and file access) may look like an easy choice for the infrastructure to support your security technology. However, their use creates the necessity for planning and gathering all of the security information as part of an overall strategic plan.

Rapid or unplanned expansion of local area networks (LANs) results in quality of service and fault tolerance issues. Major changes by the IT department can cause alarms to be missed and video with latency or jittery pictures. Likewise, your security projects can adversely affect a facilities email or other administrative/business communications.

Planning of all the communication technologies available is important to consider in the overall network utilization. Keep in mind that the topology planning of a network can contain many nodes (input points), each of a different type to effect a combined usage solution. These considerations are best discussed with a network expert that understands your specific application and the use requirements of your plan.

New and emerging technologies allow configuration, buffering, storage, and remote storage to be balanced with network requirements and load. These technologies can help to insure that the essential functions of the network are never overloaded.

Some of the solutions in an overall LAN/WAN configuration might include:

- Traditional LAN infrastructure of CAT 5e/6 with routers and switches
- External WLAN points brought to a central control (Command Control) for a District or University environment.
- External Storage and backup locations
- Subscription based access and storage

In general, we try not to rely solely on the primary network. When your primary (business) network was installed, there is a good chance that spare fiber or copper was installed for expansion. Where possible, utilize these spare cables to build a physically separate security network. Where no spare cable exists, it may still be a good idea to install some new cable to dedicate to balance the usage of the security network.

This provides two benefits. The first is to isolate high bandwidth demands like video streams and automated backups. The second benefit is a separate network keeps the everyday business users away from the security systems. While physical and electronic separation is an ideal solution, it is not always the most cost effective way to handle the requirements of security systems on an educational network.

Wired versus Wireless Infrastructure

Wherever possible we prefer to use wired networks. Wired networks are a proven infrastructure medium that is generally easy to secure and manage.

As consumers, we are definitely moving toward a wireless society. Wireless network connectivity has expanded from predominately onsite to city wide and larger geographic coverage. We know of wireless networks comprised of multisite installations pushing video, voice, and data over 20 plus miles. The ability to rely on these types of solutions solves expense of leased lines and the logistics problems involving permitting, trenching, and providing your own long distance cable runs.

Applications of these types require the ability to perform the necessary surveys to ensure that the engineered solution will work well. A tree canopy and topography survey to ensure line of sight is the first step. Due to the proliferation of wireless products all operating in the same or adjacent frequency spectrum, a spectrum analysis survey may be necessary to determine possible interference and existing traffic. Do not assume everything will work flawlessly because it is your campus or your licensed frequency (if you chose that path). Lastly, some products require licensing of the frequency over the path, which can be time consuming.

Authorized Users/System Access

Using your facility's network as the infrastructure for your security systems is a practical solution; however, access to online security related data is a concern. For that reason, not everyone should have access to your security systems.

All networks should be designed and implemented by experts having the specific network or wireless network experience. Your IT department needs to be heavily involved as network technology is changing as fast as the previously discussed security technologies. You will want to have the latest offerings/solutions information available as you finalize your network plans.

As always, the safety, reliability, and security of the networks are of primary concern, all significant/bullet-proof security precautions need to be employed. There have been instances where students have "hacked" in and taken over video surveillance systems, so careful consideration should be taken regarding wireless or wired networks, and whether to physically, or logically isolate security systems on the network.

The creation of a permissions group is a major first step in securing who has access to your systems. These permission groups create a hierarchy for user access, monitoring, control, and overall configuration of your physical security systems. Additional consideration should be made for securing the network devices that make up the network your systems will be connecting to.

Securing the information traveling on the network can also be accomplished by other IT methodologies such as VPN's, secure routing, switch port isolation, and the like. These methodologies create a separate and protected network within a network.

In larger systems that are more complex, firewalls can play a major part allowing only traffic and connections required by the security system and its necessary connection points across the network or into a specific subnet.

Typically, the programming of firewalls and Layer 3 switches to this degree requires a greater degree of IT knowledge than is found with a security integrator. It is important that this network component be identified in your process and budgeting. The programming and isolation of the network can be a costly addition if overlooked, and become a cost overrun later in the project.

Hybrid vs. Digital

Purely digital (aka network or IP) cameras that plug directly into the network are becoming less expensive, but they are big users of bandwidth because they have to stream video on the network all the time to the recorder (not just when someone is viewing the camera).

If your primary network is not robust enough to handle the new security video loads (or your IT department won't allow it, then you can build a separate network or a hybrid system. However, a separate network adds costly hardware.

Another option is to install a hybrid system where analog (not digital IP) cameras are wired directly to DVRs, and only those DVRs are connected to your network. In both of these options, security video or security related data only travels on your primary network when an operator needs it.

These considerations also apply to access control and fire systems, but the amount of data that needs to travel on a network is much less.

Security System Servers

Security and life safety systems operate 24/7/365. To ensure minimal down time, security systems should use servers and network infrastructure that are built for this environment. IT directors have learned to use special "server" class computers and "enterprise class" network components for mission critical services like email and centralized file storage.

Server class and Enterprise class components are designed to deliver high availability with a low Mean Time Between Failure (MBTF). Server class and enterprise class components are more expensive, and optional redundant / "hot standby" system designs can drastically increase the equipment needed.

Be aware that many security manufacturers offer components and computers, which are entry-level performance in order to be price competitive. These entry level systems are not necessarily bad products, but they are generally inadequate for any long-term physical security technology application.

Purpose built security servers are now available from value added computer suppliers. These high-performance servers are built by computer suppliers who work closely with security manufacturers (access control, video and others), and are often certified to work with specific security software packages.

Since cost is always a concern for schools, you will need to be selective with what equipment you choose. Depending on the importance of a particular system, you may be satisfied with a system that does not have the fault tolerance of a higher priced system.

However, do not make this choice too hastily; remember that students will be accustomed to instant access (like their dorm, computer lab, athletic areas, etc.), and staff will appreciate not being locked out of the school (break rooms, classrooms, supply closets, etc.). If the access control system fails in the case of a dorm room, you may have to station security guards there until it is fixed. In the end, utilizing non-fault tolerant equipment usually results in a greater liability and higher cost you more than, if you had planned on the more expensive servers and network components from the beginning.

Servers are frequently rack mounted with other electrical equipment. All this equipment in one place needs to be powered and generates heat. Part of your planning should be to insure that any spaces where you will install equipment are air conditioned and powered appropriately (discussed in chapter 5).

Additionally locating all your equipment in one place can be devastating should there be a catastrophic event (flood, building collapse, fire / sprinkler damage, etc.). Backup power and redundant infrastructure connections should also be considered.

Cloud Computing & Cloud Storage

Online cloud computing (where the actual programs reside offsite) and online storage solutions have entered the security marketplace. These developments have proven to be highly successful and cost effective for consumers, but have not been rapidly embraced by security applications due to concerns over security of data, potential loss of data, and business continuity should the cloud provider become inaccessible.

At the printing of this book, a leader in consumer cloud storage is reported as having an application wide security incident due to employee error. The potential damage from this breach is unknown, and is potentially significant because the cloud storage provider does not encrypt data prior to storing it offsite.

With that said, we do recommend the use of these services as part of your Emergency Recovery Plan provided the service encrypts your data prior to transmission to their secure cloud. This is not the same as an encrypted transmission where the transport of the data is protected from eavesdropping. With this type of protection, critical documents and backups of your security systems' configuration files and databases can be safely transferred to cloud storage.

Cloud storage involves data storage at offsite data centers usually with the aid of synchronization software. This software automatically manages the storage and retrieval process. At first glance, online cloud storage options with very low cost per gigabyte appear very enticing for video surveillance storage. Several options have become available for "private cloud" backups, the most viable of which allow you to own and manage the server where your security data is stored.

At the writing of this book several technical issues exist which keep us from recommending any form of cloud storage for ***streamed video recording***. Constantly streaming video creates high bandwidth demands, which translates into an internet connection that is too costly for most school environments. The drawback to most if not all cloud synchronization software packages is the limitations on file size and synchronization processes that are incompatible or too slow for this application.

We expect to see continued advances in these areas. If you are interested in pursuing one of these technologies, we suggest a rigorous pilot project to prove the concept will work in your environment prior to purchase.

Include Your IT Director

Your IT director can be an important resource in this part of your project because you will most likely be using or connecting to his network. IT directors can help you select servers and network components that your IT department can support in-house. Failing to recognize this coordination task can result in a separate maintenance cost to support non-compliant components.

Your IT director can also be a roadblock if you try to attach security equipment to your facility's existing network without coordination. It is his job to safeguard the network and keep IT related systems running.

CHAPTER 7 – ANTICIPATED MAINTENANCE AND INSPECTIONS

Maintenance

Maintenance needs to be included in your annual budget to keep your security investment operational. The way your project is purchased (negotiated or bid) is relevant to some of your options. If you already have equipment, then you may have to do some investigating to determine if your equipment has an active warranty. In this case, extended warranty coverage is frequently negotiable through the integrator (Installation Company) who initially installed the equipment.

Security product manufacturers usually include equipment warranties of 90 days to 12 months. Some equipment warranties have verbiage indicating "lifetime warranty". In this case, read the fine print. Lifetime is the device's anticipated lifetime (as anticipated by the manufacturer).

You should also know that warranty start dates vary between manufacturers. In some cases, the warranty will start from date of purchase. In other cases, the warranty will start with beneficial use (beneficial use described in chapter 10).

A labor warranty is separate from the equipment warranty and covers the cost to remove defective equipment, ship it to the manufacturer for repair, and then reinstall and configure the repaired equipment.

We recommend that you require at least a one (1) year, parts & labor warranty from the integrator and manufacturer from date of installation. It is the responsibility of the integrator (installer) to purchase additional coverage or self-warranty equipment when this duration is outside the manufacturer's standard (free) equipment warranty. In either case, get the warranty in writing.

When installing a new system, ask respondents to quote second and third year warranties in their bid (this sometimes reduces the cost over purchasing them later). Integrators obviously want to make money on all contracts, but maintenance contracts can be the most profitable if managed correctly. Integrators relying on the maintenance income will generally improve on their overall bid price.

Another option is for your institution to self-maintain. An increasing number of schools, universities, and institutions have their own maintenance staff. Self-maintaining every aspect of your security system may not be realistic due to licensing, regulatory requirements, and training. However, there is no reason that some of the maintenance cannot be performed by your maintenance & IT departments.

Traditionally, manufacturers only warrant installations by factory trained and licensed security integration companies. Today, manufacturers are increasingly offering factory training that is also available to owners. Training your staff to service the installation can be added as a requirement of your security bid.

Another maintenance consideration is documentation. All of the documentation for any installation should be retained for future resolution of problems (Your O&M documents). Installation, operations, and programming manuals are essential to the long term care of your installations as are the "as built drawings" showing the exact location of all devices and the wiring diagrams. Most manufacturers have end users support websites or email subscription services to keep you up to date on changes, software and equipment upgrades, new technology options for their equipment, and technical support.

Inspections

In many states, inspections are required by law, especially on fire alarm systems. Whether required or not, all security systems should be inspected on an established schedule as a routine part of all maintenance programs. If you use a central station or monitoring center, require that they send you a copy of their inspections for your records.

Fire systems

Most fire systems are required to be inspected on a quarterly basis. You or your contracted inspection company should notify the central station or monitoring center that you will be testing a specific system. Insure that the system is operating normally and activate the systems. Check with the central station or monitoring center to determine exactly what signals they received and insure that they were appropriate for the actions taken. Notify the central station or monitoring center when you are through with the inspection and get the name of the individual or operator number for your records.

On an annual basis, you should not only test the required functionality of the system but also perform a visual inspection of all components (VIP-Visual Inspection Process). This insures that all devices are operational and free from dirt, dust, or debris that could impair their performance or cause them to fail at a critical time.

Physical Security Systems

A bi-annual inspection and review of all operational components is recommended to provide much of the same benefits that we note under the fire system inspections. Sometime small problems corrected early can save a large expense or critical failure when your system is needed most. When inspecting specific systems each system should be visually and electronically inspected for proper operation. All devices should be individually tested to insure that they are still recognized and fully functional. A record of testing and maintenance performed should be maintained for comparison over the life of the system. A record of these quarterly, biannual, and annual inspections need to be meticulously maintained for warranty and liability reasons.

This is the time to fulfill any agency or state requirements to inspect and test your security systems. Inspections are not a waste of time, as they will frequently identify problems that go undetected in normal operations but are potentially dangerous and could result in undetected intrusions, fires, loss of system use, or costly repairs. The additional benefit you gain as your system matures and you begin to update systems is that you have a record of what has performed well. The performance knowledge give you the opportunity to address items that may have been sub-par or identify a manufacturer whose equipment has been exemplary.

Reoccurring Licensing as a Long-term Maintenance Cost

As you research the products that will become the basis for your security systems, you should take note of associated reoccurring licensing or maintenance costs. Licensing per user, per channel, per seat (PC), per interface licensing by some description are common in today's video surveillance (VMS), access control, and PSIM products. To the novice, this practice might appear to be an unusual burden, but there is a reasonable explanation for why it exists and will probably continue to exist for some time.

Consider the example, of a Video Management System (VMS) that can use open-source IP cameras. Using a particular system, your users can view, record, playback from thousands of cameras from dozens of available camera models. Because the VMS manufacturer does not sell the cameras, or the PCs you run their system on, they have to find another way to fund research and development to continue to grow and build features.

The reason for the incremental camera charges also makes sense when you consider that an institution installing a 10 camera system does not want to pay for a 10,000 camera VMS that they may never grow into. Similarly, user and PC seat licenses help to balance the cost of supporting a robust & scalable system that you can grow into.

These incremental licensing costs translate into annual software maintenance fees after the first year. For most manufacturers, software maintenance only covers bug fixes within the version you currently own. Other manufacturers include version upgrades in this service.

Software maintenance fees are for the software only, so make sure service agreements also include the labor to apply the associated software patches and upgrades. Lastly, some manufacturers require continuous software maintenance coverage if you want to upgrade your software later on.

CHAPTER 8 – DECIDING WHICH EQUIPMENT TO USE AND DEVELOPING YOUR BUDGET

The selection of security equipment is a determination of lasting consequence. Like marriage, it is not to be entered into lightly or unadvisedly. For the purposes of this book as a primer, we are discussing Security, Fire, CCTV, and Access Control.

In some cases, you may have legacy equipment already onsite, and while the initial reaction is to expand the existing system, this may not be fiscally viable or meet your long-term security & life safety requirements.

Research is the key to deciding which products are right for your project. If you are new to the security industry, you can start your research with manufacturers' websites for the type of products you need. However, nothing beats first-hand knowledge. We highly recommend that you and your team attend demonstrations of products that are representative of your intended application.

All stakeholders should attend these demonstrations and each vendor should be evaluated as to how they may fit into the institution's overall security goals. To educate stakeholders without tainting their expectations, you (or your consultant) need to facilitate more than one product demonstration.

These demonstrations can be visits to other facilities to see what they use. Another option is to contact manufacturers directly and arrange for a demonstration. The primary concern we want to point out with manufacturers' contacts is they will have an inherent desire for you to buy their products. Manufacturers may only have a limited industry experience or their experience may be limited to just one type of security technology.

We give you a head start on where to begin your research in Appendix A.

Developing Your Budget

With budget constraints, few if any institutions have the resources to totally embrace all their security issues in a manner consistent with the need. Although money is not always the answer, it is a component of the solution and must be planned and applied prudently.

During the course of identifying equipment and considering all the variables relevant to your project, you will have started to build a financial picture of what it will cost to install your system.

In previous sections, we presented a comprehensive overview of relevant systems and the components that make up those systems. Below, we outline rules of thumb, and point out some of the less obvious considerations and ancillary costs.

License Fees and Software Maintenance

License fees and software maintenance fees we describe in the previous chapter can be a very unwelcome surprise if you do not recognize these reoccurring costs in your long-term budget.

IT Related Costs

Most security technology projects will include IT related components, such as antivirus technology, system patches, OS upgrades, database management, backup and archiving, increasing network bandwidth, and network hardware needed to insure quality of service (QoS).

Many of these may seem like they would be someone else's problem, but your project may be forced to cover your percentage of some IT related resources. For example, if your system uses the Principal's desktop computer to access security systems, that computer needs to have a current and compatible antivirus. Security manufacturers do not have the resources to test their software against every antivirus product.

If there are shared resources with IT, your project coordination should include a Memorandum of Understanding (MOU) to identify shared responsibilities and understandings. We talk more about the importance of MOU's in chapter 10.

Spares

Many facilities assume the installation/maintenance company has spare equipment on the shelf. The reality is that most installation companies only offer "Depot Service". Depot Service is where they diagnose the problem and send the bad component back to the manufacturer for repair. There is usually no guaranteed turnaround time for repairs, so that portion of your system for the duration.

The solution is to purchase anticipated spares (i.e. camera, card reader, smoke detector, etc.). For components, we recommend you purchase five percent more of any peripheral device that will total twenty or more. For example, if your application requires twenty smoke detectors, buy at least one extra to use while another is repaired.

Remember that you will probably have several different models of an item (such as cameras) and will need spare stock of each type. While not exactly a spare, we recommend redundant and fault tolerant servers for mission critical systems when you can afford it.

These spares can be stored for you by the maintenance company, but we usually recommend they be stored onsite to insure they are always available. There may be some items that your maintenance staff can be trained to install.

When considered together, the installation costs, reoccurring maintenance/inspection costs, and reoccurring software licensing/maintenance costs start to bring the Total Cost of Ownership into focus.

CHAPTER 9 – CONTRACTING VEHICLES FOR SECURITY TECHNOLOGY PROJECTS

There are so many variations between different states and localities that it is impossible for us to definitively address every nuance of solicitation type or contract vehicle. It is up to the individual school, university, or institution to stay adequately informed of the guidelines regarding the procurement process for your institution. While we will be speaking generically in this chapter, our comments reflect years of experience observing and participating in all aspects of the acquisition process as a bidder, specifier, consultant, and buyer.

Due to the financial size of most security projects, your purchasing department will almost certainly require a bid process. Purchasing departments primarily deal with commodity products like paper and cleaning supplies, which conform well to the lowest bid process. Security technology projects are not commodities and the lowest bid process (even with bonding) does not protect your timeline or guarantee a successful project.

Engineered Bid Projects (aka Design-Bid-Build)

Engineered Bid Projects are projects where the complete technical design is prepared as a construction document which bidders rely on to prepare their bids and build the project. These are also referred to as Design-Bid-Build projects. Under this type of contract method, projects are usually awarded to the lowest bidder. With all bidders competing to win the project, the assumed benefit is that you get the lowest possible price for your project. The respondents are bidding and building from a design that is presumed to be 100% complete with all contingencies covered with allowances.

Any ambiguity in the specification creates delays and costs (change orders). The adage applies, "lowest price does not guarantee a quality project".

Often we see verbiage where the buyer claims the right to select the bid that they deem most favorable to their school, university, or institution (referred to as "best value" determinations). Having bids protested due to not selecting lowest bidder creates a difficult situation with possible suppliers on future solicitations. It also creates an atmosphere of us against them instead of a mutually cooperative effort, which most often produces the best result.

Be warned, we have seen instances where the low bidder, if not awarded the bid, may seek to take legal action or, at the very least, threaten to appear before the school board or other governing body. Where "best value" determination is used, we recommend your process be as transparent as possible to back up your decision.

RFP Request for Proposal / RFQ Request for Quotation (aka Design-Build)

Request for Proposal and Request for Quotation are viable Design-Build options which give you more control when it comes selecting your supplier but less control over the system design. The idea behind this contract vehicle is to tell suppliers what functionally you need and where, but let them deliver bids with equipment they are familiar and comfortable with.

You then evaluate each respondents bid individually with a formal grading criteria and choose the respondent with the best overall grade (best value). Your grading criteria should define "meets" as well as include weights for certifications, training and references relevant to your project. Just like in the design-build process, you can also pre-qualify bidders before you deliver project documents.

Under the Engineered Bid (design-bid-build) process we described above, you can expect a lot of upfront development and reasonable project management during installation. Under the RFP/RFQ design-build processes, you can engage vendors with your project faster and with less upfront development, but you will be faced with considerable coordination and management efforts throughout the project.

We cannot stress this enough, the design-build process requires individuals with the right type of skills and technical background and who are able to dedicate considerable time during the entire project. We have seen expensive change orders occur, and projects "fall off the rails", because a facility's decision makers were pressed for time and made technology decisions without performing their own due diligence.

Negotiated Projects

In some projects, you may be allowed to negotiate with contractors to get the best value for your project without a lengthy bid process. Negotiated procurement is another type of Design-Build process, which often sometimes uses a formal Request for Quote (RFQ) contract vehicle.

In a negotiated procurement process, you still receive bids from contractors, but you are not required to award a contract. The drawback to this type of procurement process is you may receive a lower number of contractor responses because your commitment or funding is unknown.

Improving on Quality of Responses

As noted, the lowest bid requirement can have negative effects on your project. You can improve your chances of getting a quality installation by using "Prequalified Bidders" criteria and/or disqualifying respondents "Outside the Median".

Prequalified Bidders

Prequalified bidders (prequal) are an easy concept to understand, respondents are required to meet some project related criteria. This already occurs to some extent on most projects where respondents are required to provide proof of insurance, proper licensing and employee safety policies.

We recommend prequalifying bidders on all your Physical Security / Life Safety projects so you do not expose your security plan to the world (specific building plans and your intentions on how to secure it). Some of the qualifications we recommend are described in the Chapter 10, "Developing Your Bid Specification".

Discard Outside the Median

"Discard Outside the Median" is a bit harder to swallow by some purchasing directors because you may end up discarding the lowest bidder. A properly investigated and well-designed "design-bid-build" project will anticipate a majority of the technical hurdles and significant expenditures. Public bids with high quality design and comprehensive specifications will usually result in bids that are within 10-15% of each other.

Respondents, whose bid prices are vastly higher or lower than the median of bids, have usually missed some critical details or components. Including a "discard outside the median" clause, allows you to reject these bids as "non responsive due to defect". This type of clause forces the respondents to give their bid response the attention it deserves and not shortcut the estimation process.

On one of our projects, the bid documents required 30 days of video storage but one respondent anticipated only 14 days. Because this project had a large number of cameras, this oversight represented several hundred thousand dollars. Using this clause, we rejected the lowest bidder and awarded it to the next lowest bidder who was within 10% of the median. Had we accepted the lowest bidder, the respondent may not have been able to finish the project or may have performed lower quality work in order to protect his margin.

We recommend this protection clause when a high number of respondents are expected and your project is a high dollar project.

Best and Final Offer

All contract vehicles can include a "Best and Final" final review where bidders whose bids are within close range of one another are given the opportunity to adjust their bids based upon discussions and clarifications of requirements. As the name implies this process allows bidders to revise their bids (up or down). You will need to let bidders know that this will occur and if some or all bidders will be allowed to participate.

By using Best and Final Offer, Bidders (who would have been discarded solely on their bid price) are given a second chance to address missing costs or miscalculations.

CHAPTER 10 - DEVELOPING YOUR BID SPECIFICATION

In the previous chapter, we talked about the bid process and potential requirements that we recommend to make your project more successful. In this chapter, we will expand on those considerations and offer additional recommendations for developing your physical security bid specifications.

One of the most critical tasks in your project is developing the bid specification. The specification communicates your needs to bidders and often becomes part of the contract document. Omitting information or incorrectly identifying site conditions are two of the most common reasons for change orders and project conflicts. Consultants, engineers, and architects carry Errors and Omissions insurance to protect against such occasions.

To insure your project succeeds, your specification needs to be well documented and have a clear definition of the goals you wish to achieve. Below we introduce you to some of the critical pieces.

Specification Formats and Standards

Most consultants and engineers use standard formats and sections for bid specifications, which were established by either the American Institute of Architects (AIA) or Construction Specifications Institute (CSI).

Until 2004, Physical Security/Life Safety and IT projects were lumped under AIA electrical division 13, AIA special systems division 16, or CSI specialty systems division 13. In addition to being confusing, the result was specifications which left the owner exposed to change orders and electrical contractors, who were not familiar with the nuances of these technologies becoming the prime bidder on the project. Electrical contractors treated products in these sections like brick-and-mortar commodity projects and often passed work through the

lowest bid subcontractor irrespective of their qualifications.

In 2004, Security and IT technology projects were identified as having different needs and skill sets. At that time, CSI released its MasterFormat™[3] standards, which identified security technology projects as having its own division (28 – Electronic Safety and Security).

---

[3] MasterFormat is the registered trademark of the Construction Specifications Institute. Additional information about CSI is available in Appendix A.

The significance of this change was the recognition of a highly technical field requiring a different contractor with specialized skills. Today it is common (and rightly so) for security installation companies to be the prime contractor on a Physical Security/Life Safety projects.

As a component of a successful project, we highly recommend that your specifications follow the current CSI MasterFormat™, which has become the de facto standard. By following this standard, you will be speaking the language of architects, engineers, and specifiers, which will help contractors, and suppliers meet your requirements, timelines, and budgets. While the CSI standards and formats originated to support fully engineered design-bid-build projects, they are also relevant to design-build (RFP & RFQ) project specifications.

Product Specifications

One of the most important tasks you will do in preparation for bid day is to accurately describe your project's requirements. Consultants spend a great deal of time articulating the intended use, installation, and maintenance requirements for each system or major component.

In a ***design-bid-build*** project, your specification is expected to be substantially complete from an engineering design point of view. The product specifications you include here will specify technical attributes that respondents will be required to deliver. This document is referred to as the "Technical Specs". Products will be described in detail including dimensions, construction material, expected power use, physical connection points, and in some cases actual makes and models.

For example, part of a technical spec may include, "The network video recorder shall record video using the video compression standard H.264 and shall include 5 terabytes of Raid 5 fault tolerant storage with hot swap hard drive storage trays accessible from the front of the unit. Hard drives shall have an MTBF of 1 million hours".

In this example, bidders know exactly what you want. You determined that H.264 video compression has the quality and the required storage footprint you want. You also anticipated aspects to protect against outages and considered ease of changing any failed hard drives. Because you indicated the amount of storage, respondents do not have to concern themselves with proposing something you may or may not want.

In a ***design-build*** project, your specification will focus less on the technical specification and more on the required technical functionality and specific standards. These "Functional Specs" give bidders much more flexibility to provide products they prefer.

For example, part of a functional spec may indicate, "The network video recorder shall record video continuously for 14 days on RAID 5 protected hard drive storage using H.264 compression at 15 frames per second with 640x480 resolution".

In this case, the respondent can deliver equipment that they are comfortable with, but they are responsible for calculating and providing storage for 14 continuous days of recording. The requirement for H.264 video compression also serves to narrow down the products that can be submitted on by bidders, but allows multiple products to be bid.

This can be both good and bad. The functional spec example also allows respondents to deliver the video storage component any way they want. For example, respondents could deliver the storage with no fault tolerant functionality. They could deliver storage internally or externally.

Regardless of specification type, it is up to you to make sure that your wants/needs are clearly communicated. If your specification is not clear, you may receive bid responses that meet the letter of the specification but not the intent.

Commercial Off-The-Shelf (COTS) and "Or Equal"

To avoid the pitfalls of specifying a product that cannot be competitively bid, you will want your specification to describe characteristics of commercial off-the-shelf (COTS) products. COTS products are products, which manufacturers actively sell, support, and maintain in a commercial setting. By referencing attributes of a known COTS product, respondents will know you need something that already exists (not a custom developed product or system).[4]

As you list required attributes, remember 'Features' are "nice to haves" and 'Capabilities' are "must haves". If you discover "nice to have" features that you do not need immediately but feel will be helpful in the future, include them as a biddable option. This way you can get as many competitive bids as possible and review the bid options with regard to best overall value.

---

[4] Products are considered Custom (non-COTS) when any part is created or permanently modified only for a specific project. Custom products are not covered by this book because they are universally more expensive and future maintenance/upgrades are considerably more troublesome than COTS products. The exception to our moratorium on custom products is when items are customized for fit and finish which is easily reproduced in the future (such as custom paint).

With that said, you need to carefully select the equipment attributes required and not mix attributes from multiple COTS products. Mixing system/component attributes creates an "unbuildable" specification also known as a "pie in the sky spec". Consultants and engineering firms who are not Physical Security / Life Safety subject matter experts have been observed to make this mistake often. The unfortunate result of this mistake is often expensive change orders after a bid is awarded.

In most cases, your specification will not direct respondents to provide a "single (sole) source" solution or product (described later in this section). Your efforts in educating yourself and investigating products will benefit your specification and system design criteria. Usually, your efforts will be tempered by the purchasing department's requirement for competitive bids. Most purchasing departments will require preferred products or product capabilities referencing a specific product are suffixed with "or Equal".

Integrators who intend on submitting an "Or Equal" solution or product should be required to prove that it is, in fact, "Equal". Integrators prove this by submitting information substantiating "equality" within a prescribed time so it does not draw out your bid process. "Or Equal" proposals need to be handled immediately, so your process cannot be faulted as unfair. Good prior research will insure that you have already evaluated the equipment and decided for/against its use. Should you be presented with something you are not aware of and consider it worth investigation, you should delay the bid submission date until you can properly investigate it for consideration.

Your approval of any "Or Equal" submission should be made known to all bidders for fair consideration in their responses. "Or Equal" approvals should be identified as conditional with the requirement that a comprehensive demonstration and final acceptance are required prior to receiving a Notice to Proceed. A statement like this should be included in the original specification, so this process can be enforced without protest.

Sole Source Justification

Sole source justifications have their place in the overall scheme of procurement.

In the Physical Security / Life Safety industry, products between manufacturers are not generally interchangeable. In these cases, sole source is easily justified when your project is expanding an existing system, whereby you need to maintain product from a particular manufacturer.

Aesthetics can be another reason for sole source justification where new equipment will be placed around existing equipment or selected to fit in a specific environment (such as historical sites or hazardous areas (chemistry labs)).

Lastly, there are applications where a product contains patented content or functions not present in any other product and that feature set is required by the project.

Sole source justification requires a value judgment and cost analysis when equipment has reached or approaches the end of its life expectancy. In this case, it may be more reasonable to replace the older system rather than continue to support it as a sole source product.

"Leading Edge", "Bleeding Edge", and Pilot Projects

If your sole source justification refers to ***new*** technology, there may be reason for concern. New is a relative term with different interpretations depending on the user and carries with it different connotations and expectations. New security technology is usually described as either "Leading Edge" or by some, "Bleeding Edge" to convey an associated degree of risk and expense.

"Bleeding Edge" products have a high risk of being unreliable and early adopters are likely to incur high expense implementing them. Successful implementations and testing are usually lab based under controlled environments. If these products are available from multiple manufacturers, they may be significantly different.

"Bleeding Edge" products are notorious for underwhelming an owner. Beware the sales representative who can only demonstrate his product in PowerPoint or has no installed customer base!

Every year numerous manufacturers and products enter the security industry only to be gone in a year or two. Some of these new startups are last year's failure under a new name. It is not uncommon to see 100+ new manufacturers exhibiting at a tradeshow from year to year.

Conversely, "Leading Edge" products are those where consensus is evident between manufacturers and successful implementations can be referenced from real life customers.

This is not intended to be a moratorium on new and innovative technology. Consider that at some point, electronic mail (Email) was once considered bleeding edge, and then it matured into leading edge, and finally became mainstream.

Rather than falling victim and wasting a lot of money, consider running a Pilot Project to prove out the technology. Your project can include a pilot project with the option to continue or expand. To insure the highest number of responses, be sure to identify if you have the funds to complete the pilot and full installation if desired.

"End of Life" Products

Just as we caution you regarding first run "bleeding edge" products, we also caution against regarding buying the last product off the assembly line. This recommendation is not universal. We are not concerned with common products like wire or computer keyboards. We are talking about any product that is critical to your system or your aesthetic setting.

For example, fire pull stations and security cameras are fairly common and easy to source. However, you probably want all of them to match (fit & finish) in the same hallway or the same room. This is a simple example, but "end of life" can also affect functionality you rely upon.

Manufacturers introduce and discontinue product lines all the time for various reasons. You should be especially careful when a manufacturer acquires another manufacturer. If the acquisition creates a situation where the parent company now owns two competing product lines, one product line will be discontinued at some point or the two will merge. Until it is resolved, the manufacturer will claim they are "supporting both products for the foreseeable future". From experience, "foreseeable future" is anywhere from six to 24 months.

When evaluating products, request in writing how long the manufacturer will produce and maintain specific models and versions. If your concern is strictly aesthetics, then your best course is usually to buy a stock of spares.

Bidder Qualifications

The following bidder qualifications are not all encompassing. Many consultants and purchasing directors will have their own preferences. The qualifications we include below are intended to avoid many of the pitfalls that can threaten a project. These recommendations apply to both prequal and non-prequal bidding processes, and are in addition to the standard due diligence checks (financial standing, work history, references, etc.).

Local Service Presence

Security providers that have an established and financially secure base in your area will normally be your best option. This local presence helps to reduce response times and maintenance duration due to their proximity of personnel and repair stock.

We recommend requiring the installing/servicing company maintain an office and personnel within sixty miles of the school or university campus. Be wary of respondents who promise to open an office if they win the project and put in verbiage allowing you to remove them from the project if they do not provide it. With today's internet based phone services and mail forwarding/hosting providers, your specification should define what constitutes a local office.

Depending on the needs of your facility, it may be more advantageous to house repair stock onsite and give service personnel office/shop space.

Proof of Certification and Experience

If you had not already guessed, physical security technology has quite a few nuances and technical specialties. Your specification should require respondents to prove they are certified under various standards that apply to your project and locale. Likewise, they should require the respondent to be trained and (if offered by the manufacturer) certified for the products they intend to provide.

Make sure the language of your specification requires that the certifications presented by a respondent are for the individuals who will be onsite performing work. Your specification should also require the respondent maintain those certified persons and commit to replacing them with another equally certified person should the original person(s) leave.

To be competitive, you may allow respondents to propose products that are new or new to them. Your specification should require the respondent to demonstrate that submitted product manufacturers have performed similar work of size, scope, and nature as well as being trained prior to commencing any work.

Teaming

Another bidder consideration is teaming. Teaming is a great way to bring specific expertise to a project. For example on design-build projects (RFP & RFO), security consultants can be relied upon during the design and oversight portions of the project. We have also seen teaming responses where one company will provide the installation and project management, and another will provide local onsite service after installation.

Contractually teaming arraignments are really a contractor / subcontractor relationship. Your specification should identify if you allow teaming and what additional qualifications the respondents are required to meet.

Testing & Acceptance

Often your procurement or engineering department will have some standard specifications or requirements for Testing and Acceptance. These specifications will usually make a good starting point for testing criteria on your project. They will generally need some modification because most are created for construction projects, whereas Physical Security / Life Safety projects have aspects of both construction and technology.

Make sure you have all stakeholders involved in the planning for this project. We have seen projects delayed for 12 months or more because testing procedures and roles of acceptance witnesses were unclear. It is a good practice to put a Memorandum of Understanding (MoU) into place for all involved departments.

It is important that this MoU be in place for all parties that will have a hand in testing approvals or maintaining any aspect of your project. For example, get your IT Director to outline what he needs in order to take on the responsibility for supporting the increased network infrastructure demands.

In addition to outlining test procedures and the roles of acceptance witnesses, testing specifications should include a clear description of what should happen if testing fails. Testing is supposed to find problems. You cannot expect testing to pass 100% the first time, but neither should you expect to retest repeatedly.

Stakeholders will sometimes stall testing indefinitely by not approving a test because they do not understand the purpose of a test, or if they decide during the test that they want/need something different. Make sure your testing requirements provide a well-defined test procedure, process for correct measures for system defects, and prescribed retesting process.

Your specification should also indicate who has oversight and final approval capability on the project.

Once you award the project to a respondent, one of the things they will need to deliver to you is a testing plan addressing all of the testing requirements you have specified.

Proof of Concept Mockup (Optional)

Your project can include the requirement for a proof of concept, or ***mockup***, demonstration prior to issuing the Notice to Proceed (NTP) with complete installation. This practice is useful when you have concerns regarding how new equipment may interact with existing equipment, or if the project will interface mainstream products in a non-standard way.

Earlier we described a pilot project, where you pay to try a new technology separate from other work. In this case, proceeding with the entire project is contingent on the success of the mockup. This practice helps to flush out unanticipated problems before an installer starts messing around onsite.

Mockups should be a reasonable demonstration of the questioned components or questionable combination of components. It should be comprehensive but should not waste time and money (yours or the respondent's) by testing every piece of equipment.

For example, there is a good chance that your new project will rely on your facility's network infrastructure in some way. However, do not force the respondent to recreate your entire network structure unless necessary (network equipment is built on standards and is predictable/testable). A more realistic mockup can be coordinated with the IT director to occur onsite using your existing network after hours or between school sessions.

Pilot projects, mockup testing, and approval criteria should be well documented. Failures during mockup testing may uncover an inexperienced installation company or it may uncover that your approach or existing resources needs to be improved. This may halt the project before excessive resources have been expended, or a change order can be initiated out of contingency budget.

FAT Testing

Prior to equipment installation in the field, equipment should be sampled under a Factory Acceptance Test (FAT). In FAT, you will test relevant equipment to insure that it meets the specifications indicated by the manufacturer. Less mature COTS products should be required to go through some additional load/capacity testing to identify issues before you get the whole system installed.

System Testing

Whether your project will be installing one system or multiple systems, you will need to test each manufacturer independently prior to completing any combined system testing. Each system type will have testing that is unique to that system. For example, cameras need to be tested to confirm operation under the video surveillance system and doors need to unlock under the access control system.

Integration (Interface) Testing

The term integration is often used incorrectly within the security industry and the companies who install these systems are often called "Integrators". This harkens back to when installation companies physically connected disparate hardware to create a functionally system. Today it is an issue of semantics, since most physical security systems are really interfacing between independent standalone products. Work defined as Integration Testing tests the implementation of manufacturers API's or SDK's that your project relies on to exchange information.

Installation of components, system head-ends (servers) and system interfacing may occur simultaneously due to the way integrated physical security projects progress. This is not always the case but common enough to mention. For that reason, System Testing may step straight into Integration Testing with no break or delay.

Integration Testing is another functionality test, which is intended to demonstrate that two or more systems are interacting with each other correctly. Chances are that your system and components specifications will be explicit and address not only capabilities you need immediately but also features that you would like or may be useful in the future. Integration between systems is going to be more defined and your testing here will probably follow the integration part of your specification.

At this point of your project, the system may or may not have programming which configures the system to work within your environment (such as users, authorizations, labels for cameras and doors, etc.).

Documentation

Your specification needs to outline documentation you will rely on throughout the life of the project and system(s). If you have never worked on a security technology project before you will be amazed at the amount of documentation needed, and how important it is. If you have previous experience with implementing technology projects, you will understand the importance and quality necessary.

Bid Responses (or Prequalification)

As part of their bid responses, or as part of prequalification responses, respondents should include documentation to prove they comply with specified requirements (such as proof of training or product certification). This is the first glimpse you will have at the capabilities and experience of the respondents.

After Award (Submittal Process)

After award, you will need to receive an installation plan from the contractor so it can be approved and you can issue the Notice to Proceed (NTP). At a minimum, the contractor should include design drawings and a work plan with proposed schedule. To insure there are no delays and no change orders for misunderstandings from your side, you will need to provide the contractor with more concrete information than the outline you provided in the bid documents regarding site conditions, existing infrastructure, existing systems, and site accessibility schedules. In multidisciplinary projects, you should also include contact information for responsible stakeholders or their designees.

Your contractor may have included product specifications with their bid responses, but we recommend that you require they provide the most current revisions with their installation plan (not marketing cut sheets) because manufacturers may have updated specifications since the bids were turned in.

We also recommend that you require Operations and Maintenance (O&M) manuals with the installation plan for critical system components and software. This is not an overly demanding request because most manufacturers are moving to digital O&M manuals and many are downloadable from their websites.

With design drawings, work plan schedule, current specifications, and O&M manuals at your disposal, you (or most likely your team) will have a better chance to identify obvious problems with the contractor's installation plan.

Project Closeout

During project installation, there will be coordination and status meetings where you give your contractor direction concerning questions he may have about fit and finish, programming/configuration, discovery of previously unknown site conditions (those little speed bumps common to all technology projects).

During this time, the contractor should be transferring those changes to the project drawings so at closeout you are presented with an accurate set of "as-built" drawings.

The contractor should also be required to make additions to the master set of O&M manuals so a complete library of all O&M manuals is turned over to you. This should be delivered to you both electronically and in paper form (due to some voluminous manuals, abbreviated paper manuals can be accepted where full manuals are available electronically).

As part of the project closeout, you should also expect the contractor to deliver training and supporting training documentation/materials as described below. These training materials should be included as part of the O&M submission.

Also as a final activity of project closeout, it is recommended that the contractor supply you with a digital backup of settings for all systems and subsystems. This will help you to recover should your system suffer a catastrophic failure.

Training

Training is another very important area you need to describe in detail in your bid specifications. If no one knows how to maintain and use your new system, then all the work you have done to get it funded and installed becomes less effective. We highlight below some of the areas for you to maintain and provide continuity for system use.

You need to consider that your system is going to have a number of different types of operators and obviously, the training will not be the same for all personnel depending on their individual involvement with each part of your security system(s).

Administrative personnel could potentially be trained in the systems management functions, assigning

codes, schedules, and getting systems data for evaluation.

Maintenance personnel could be trained in troubleshooting, testing & inspections, and repairs.

Security or police personnel could be trained in using the feature set designed to give critical information when responding to alarm situations. They will also need to learn how to maintain "Chain of Custody" under the new systems.

If you are installing an access control system, faculty may need to be trained on how to use their access control cards (often done through printed material).

Most of the specifications we see have a much-understated requirement for X hours of operator training and Y hours of administrator training. Because the specification is imprecise, the trainer can do just about anything during that time and still satisfy the specification requirement. Contractors that want your repeat business will try to do a good job, but what happens six months later when an employee leaves? Do you hire the contractor to do new hire training? That can be expensive and will probably have inconsistencies between trainings.

Your specification should be very detailed regarding what you want taught and to whom. If you desire a specific type of training (administrator, operator, and maintainer), then you will need to specifically require it. Projects that include PSIM over the top of other systems will need operator training for the PSIM and all subsystems, but you may or may not need administrator, manager, and maintainer on PSIM and all subsystems.

Typically, certificates certifying the training and course content are given to those individuals attending the training and completing the prescribed course of study. It is important that all training is documented and a copy of the issued certificate be maintained.

To resolve the issue of training new employees, we recommend your training specification also include "train the trainer". Direct the contractor to provide additional training sessions for one-on-one or small group training of persons who will be responsible for training all new hires (administrator, operator, and maintainer).

The opportunity exists where the trainer may be the employee to leave. Therefore, a good practice is to require that training be recorded and prepared as a formal training tool. This is your project so you can also require the raw video recordings and specify the production application to be used. Specifying the production application allows updating training videos to be done in house as your system grows or changes.

Final Acceptance / Beneficial Use

Most physical security technology contracts recognize the significant milestone "Beneficial Use" and the all-important "Final Acceptance". Your specification should define these two events as succinctly as possible. It is important to describe the required functionality of each event, how you will validate that this functionality is delivered, and the associated payment per milestone.

In some cases, the milestone of Beneficial Use can mean simply that the primary system or systems are installed and some functionality has been delivered. In order to receive Final Acceptance the integrator would complete all specified work. There have been other cases where the authors have seen greater demands placed on the term Beneficial Use.

Depending on the size of your project, "Beneficial Use" can span multiple locations and/or be segmented by system. Final Acceptance could be segmented by locations but all required functionality must be tested as complete on a per location basis, and as a whole.

CHAPTER 11 - WHERE TO GET HELP

Salesperson, Architect/Engineer, or Professional Security Consultant

Before you decide who will help your team through the various processes for your new system, it is important to understand the industry roles and practices with regard to the relationships of Salespeople, Architect/Engineers, Manufacturers representatives, and a Professional Security Consultant.

The **Salesperson** works for the system integrator and may be of great help in discussing potential solutions or methods for achieving your security goals. It is important to keep in mind that he has a vested interest in selling the brands his firm represents. By using a salesperson to help design your system, you may also be unknowingly allowing design restrictions that can only be met by the brands offered by the salesperson's employer (a proprietary solution). Most salespeople will offer their design services free, and attractive as that may be, it can cause conflicts in the bid process and the bid may be considered sole sourced if challenged.

An additional concern by introducing the salesperson to the project details, he now has an unfair advantage over other potential respondents who do not have prior knowledge.

Manufacturers' Representatives can be a great resource for information during your exploration and learning phase of the project. To clarify this term, there are two types of Manufacturers' representatives. There are representatives that are directly employed by the manufacturer and other representative who are employees of independent sales companies who perform sales functions for multiple manufacturers.

Factory direct representatives will generally have the strongest working knowledge of their company's products but their experience may not encompass the whole market.

Independent representatives generally have a broader industry understanding because they work with multiple product categories. Independent representatives often have experience interfacing different products to maximize their sales opportunities.

Both types of manufacturer's representatives are salespeople and present the same concerns we offer above for the integrator salesperson. Allowing these representatives to contribute to your specification can produce the same result; a proprietary specification with the risk of being non-conforming to bid standards or allow the bid to be challenged.

Architects and Engineers (A&E) are formally trained and licensed professionals that have extensive design experience and create specifications on a routine basis. As industry professionals, they have all the core requirements to build the physical aspects of your building and provide the core services you need to create bid documents.

The characteristic that most A&E firms lack is the Subject Matter Expertise/Expert (SME) with the capability to design a Physical Security project. For A&E firms, security technology system design is not bread and butter work. In the occasional approach, an A&E may be conservative and approach a project without current knowledge and lessons learned. At times, the learned A&E firms will recognize that they lack the expertise in the physical security field and will hire a Subject Matter Expert (SME) typically a Physical Security Consultant (discussed below).

Professional Security Consultants are industry professionals, most of which have experienced numerous years within the security industry. Because there is no formal schooling to learn the industry, Security Consultants have learned over time by working in different segments of the industry. These Security Consultants bring with them a wealth of experience and knowledge of equipment, projects, specifications, and the bid processes.

The question for you as a prospective client is "how do I find the right consultant for my project". We recommend that you inquire of several before selecting one individual or consulting company. Consultants' reference base and experience can be as vast as the industry itself.

It is important when deciding to work with a consultant that you understand that they can function on an as needed basis to help with specific portions of the project, or to act as the owner's representative for the project as a whole.

When deciding what level of expense and expertise you require for your project, consider the amount of resources, experience, and time needed to perform the tasks at hand.

- Risk Assessment
- Vulnerabilities & Liabilities
- Physical, Logical and Environmental Assessments
- Size / Cost of Project
- Complexity and amount of systems integration
- Needs Analysis
- Understanding Stakeholder Needs
- Including Discovery of Legacy Equipment

- Perform Trade Studies for Available Options
- Develop Project Estimates or Perform In-depth Total Cost of Ownership Review
- Specification Writing & Drawing Preparation
- Bid Process
- Handle Technical Questions at Pre-bid
- Assist with Bidder Prequalification
- "Or Equal" Substitution Review
- Construction Administration (Owner's Representative or "As Needed")
- Project Management
- Organize Project Status Meetings
- Onsite Progress Inspection
- Configuration Coordination
- Testing Oversight
- Milestone Certification
- Final Acceptance Review

Consultants by nature are agnostic to brand influences. As the owner's representative, consultants will work with you to understand the value of your legacy equipment (if you have any), what your current goals are, and what you wish to achieve with your system(s) in the future.

When selecting a consultant, thoroughly examine the consultants' references, certifications and stated experience prior to selection. Just like all subcontractors, work with a contract that includes well defined stated expectations.

APPENDIX A - ONLINE RESOURCES

We thank you for purchasing our book. We hope that it was helpful and informational.

For your convenience, the authors have created worksheets and checklists to help you with your school security projects. Additionally, we have compiled a list of online resources, training, and other information that we only briefly touched on in the book.

Access to this information is only available to persons whom have purchased this book.

Please visit http://www.schoolsecurityprimer.com. You will be asked to register for access to the information. During registration, enter the code "SCHOOLSECURITY".

APPENDIX B - ADVERTISERS

Publishing costs for the book were shared by the authors and several security product manufacturers whose advertisements appear on the following pages in alphabetical order. (The images below were originally sized to fit the print version of the book). As you can see from this list, digital video is currently a very active segment in the security industry.

- Arecont Vision - Digital Video Cameras and Encoders
- Axis Communications - Digital Video Cameras and Encoders
- Genetec - Digital Video Management
- Hochiki America - Fire Alarm Systems
- Lensec - Digital Video Management
- Salient Systems - Digital Video Management
- TimeSight Systems - Digital Video Management
- SyncBoss - Secure Cloud Storage

APPENDIX B - ADVERTISERS CONTINUED

Image quality is always important, but the benefits you get from video surveillance depend on how you use your images.

We focus on image usability, sharing our experience, competence and a full range of image-enhancing features like remote zoom and focus, megapixel and HDTV, so you can benefit fully from the industry's broadest portfolio of network video products.

To give you hands-on access to essential IP-Surveillance expertise, we partner with over 24,000 specialists worldwide. Make the most of your video surveillance capability. Get the Axis picture. Stay one step ahead.

www.axis.com/imageusability

AXIS® COMMUNICATIONS

APPENDIX B - ADVERTISERS CONTINUED

HOCHIKI
america corporation

"The world's finest fire detection systems since 1918"

Hochiki America offers a wide-variety of high-quality fire equipment products.

Safety For All - By our sophisticated life safety, fire detection and emergency devices, people's lives and property are kept protected thus helping make the world a safer place to live.

http://www.hochikiamerica.com

Simple.Scalable.Security™
Crisis Planning
Security Assessment
Emergency Preparedness
SALIENT SYSTEMS
A LEADING PROVIDER OF NETWORK FRIENDLY
VIDEO SURVEILLANCE & MANAGEMENT SYSTEMS
| OPEN ARCHITECTURE | FULLY SCALABLE SYSTEMS |
| EASY TO LEARN AND USE |
| INTEGRATION WITH ACCESS CONTROL |
| ANALOG AND IP CAMERA SUPPORT |
Salient Systems Corporation
www.salientsys.com
10801 Mo Pac Expy, Building 3, Suite 700
Austin, TX 78758
512.617.4800

TimeSight™ Systems See More.

Megapixel Images Without Mega Storage

TimeSight Systems is the industry leader in Video Lifecycle Management (VLM). An emerging best practice, VLM shrinks video files over time as they get older and less relevant, according to the rules that you and your administration decide. TimeSight Systems can store and manage high resolution video for hundreds of cameras in a "pizza box" appliance for an entire academic year, ensuring school- and campus-wide compliance with state and regulatory requirements. The result is massive savings in storage—higher video image resolution and longer retention times at a reduced cost.

This is my video storage for 30 days

Why TimeSight VLM is Significant

- **Best-in-Class Video** TimeSight allows you to use the best analog or IP cameras on the market to capture video at the highest resolution with a much smaller storage footprint.
- **Extended Storage** You can now store all video for an entire academic year, increasing visibility to threats of bullying, "slip-and-falls," and student and teacher safety.
- **More Cameras, Better Quality** By reducing your cost of storing video, you can now capture video at the highest resolution and add more cameras—with full coverage every second of every day.
- **Increased Security Profile** Better Cameras + Higher Resolution + Longer Retention = Better Security.

And this is mine for 1 academic year

TimeSight Systems, Inc.
16000 Horizon Way, Suite 800, Mount Laurel, NJ 08054
856-206-9739 | www.timesightsystems.com
Contact us at inquiries@timesightsystems.com

APPENDIX C - ABOUT THE AUTHORS

HENRY L. HOMRIGHAUS JR., PSNA, CHS-V

Henry L. Homrighaus Jr. is a career security industry veteran with a wealth of experience derived from over 40 years experience in military, law enforcement, and physical security industry involvement. During this time, Henry has led some of the largest and most successful of America's security companies to a position of prominence in their respective marketplaces.

Henry is certified as a Professional Security Network Associate (PSNA) and holds the highest certification (CHS-V) from the American Board for Certification in Homeland Security.

Henry has designed and implemented numerous physical security projects at military bases, seaports, and educational throughout the United States. Henry is the president of ProSecCon, Inc. d/b/a Professional Security Consulting providing expert witness and security technology consultation.

DIGITAL BUSINESS CARD

FRANK J. DAVIES, CIPS, CVI, CHS-III

A veteran of 30 years in the physical and electronic security industry, Frank Davies has specialized in the design and implementation of sophisticated security integration projects for the Federal Government, Airports, and Fortune 500 clientele. Frank has installed and overseen the installation of numerous physical security projects at educational facilities around the country.

He holds certifications by the American Board for Certification in Homeland Security (CHS-III) and Office of Infrastructure Preparedness. Frank is a member of several industry organizations and is a member of Infragard (FBI Private/Business Organization) and the International Association of Campus Law Enforcement Administrators (IACLEA) since 1996.

Frank is a principal of Aella Consulting Group, Inc. where he serves an expert witness and security technology consultant.

DIGITAL BUSINESS CARD

GREGORY BERNARDO, CSI, CVI, CHS-I

During his 18 years in the physical and electronic security industry, Gregory Bernardo has held key positions in sales & marketing, technical support, applications design, product management, and project management for various security industry leaders both in the manufacturing and consulting sides of the industry. During this time, Greg has designed and assisted with implementation of physical security projects at many of our country's prestigious airports, seaports, healthcare facilities, and some of the largest school districts in the central United States.

Greg has been published in the international magazine Passenger Terminal World, has been a speaker at BuilConn, and performed dozens of national and international training seminars specializing in video surveillance.

Greg is a principal of Aella Consulting Group, Inc. where he serves with Frank as an expert witness and security technology consultant.

DIGITAL BUSINESS CARD

APPENDIX D - BOOK PROCEEDS

A portion of the proceeds from this book go to support

ConsultantRegistry.Org

Serving Security, IT and Communications Consultants

Are you the missing piece?™

ConsultantRegistry.Org supports the security industry by giving consultants a place to present and showcase their skills. We also provides a starting point for technical and application research.

Find Subject Matter Experts for Your Projects

ConsultantRegistry.Org consultant members have detailed profiles and can be found in our Member Directory

Learn About Best Practices & Alternate Solutions

Perform key word searches on 100's of relevent documentsin the White Paper / Case Study Directory

Start a Discussion Group & Collaborate with like minded individuals

Technology Constantly Changes

Stay up to date with Technology Manufacturers and their products

Which manufactures have ConsultantReady tools to make Project Design easier?

Consultant Ready

ConsultantRegistry.Org